DAVID ROSWELL

ENIGMATIC HISTORIES:
Exploring Unsolved Cultural and Scientific Puzzles

«Where the enigma of history comes alive...»

2024

"The science is but the beginning; the true beautiful
world, it is infinitely deeper."

Isaac Newton

CHAPTER

INTRODUCTION

From the dawn of human civilization, mysteries have intrigued and eluded us. They stand as testament to the vastness of the unknown, the impenetrable veil that shrouds certain aspects of our existence. These enigmas, scattered across time, geography, and human experience, challenge our understanding, provoke our curiosity, and compel us to seek answers where they seem elusive.

In this exploration of the perplexing and the unexplained, we embark on a journey into the heart of enigmas—some ancient, others contemporary—each a puzzle piece in the vast tapestry of humanity's quest for comprehension. This book, "*Enigmatic Histories*", invites you to step into the realm of the inexplicable, where history, science, culture, and human curiosity intersect in a tapestry of wonder and bewilderment.

Herein lies a collection of stories that have defied explanation for centuries, puzzling scholars, scientists, and enthusiasts alike. Lost civilizations fade into myth and legend, cryptic creatures inhabit the corners of our folklore, and ancient monuments whisper secrets yet to be deciphered. The perplexities of unsolved crimes, the elusive nature of certain scientific phenomena, and the cultural mysteries that transcend time beckon us into a labyrinth of speculation and inquiry.

As we delve into the chapters that follow, we will encounter the baffling, the uncanny, and the extraordinary. From the depths of the ocean to the far reaches of space, from the foggy moors to the bustling city streets, these unsolved mysteries will enthrall and challenge our perceptions of what is known and what remains beyond our grasp.

But remember, the allure of these mysteries lies not only in their enigmatic nature but also in the shared pursuit of understanding. As we navigate through the unknown, we invite you to join us in pondering the unexplained and celebrating the curiosity that fuels our collective quest for knowledge.

So, without further ado, let us embark on this journey—a journey that will take us through the realms of lost civilizations, eerie occurrences, unexplained phenomena, and the mysteries that continue to stir the human imagination. Welcome to a world where puzzles abound, where the answers may be tantalizingly close or forever out of reach.

Let the exploration begin.

CHAPTER 1: LOST CIVILIZATIONS

In exploring the enigmatic concept of lost civilizations, we embark on a journey that delves into the mysteries surrounding legendary tales such as Atlantis, perplexing geoglyphs like the Nazca Lines, and historical enigmas such as the vanishing Roanoke Colony. These cryptic narratives and ancient engravings on the landscape beckon us to unearth the truths hidden within the remnants of these vanished societies. Join us in an exploration that traverses the boundaries of history, myth, and the enduring quest for understanding the fabled lost civilizations.

"Lost civilizations are the silent whispers of history, reminding us of the impermanence of human achievement"

Robert Ballard

THE MYSTERY OF ATLANTIS

In the quest to uncover the truth behind the legendary Atlantis, scholars have scrutinized historical records, geological data, and archaeological findings to validate Plato's accounts. While the tale of Atlantis remains tantalizing, a lack of concrete evidence has led many to view it as a literary allegory rather than a literal lost city.

Plato's descriptions of Atlantis, documented in his dialogues *"Timaeus"* and *"Critias"* paint a vivid picture of an extraordinarily prosperous and advanced civilization. According to these accounts, Atlantis was a magnificent island nation located beyond the Pillars of Hercules, often associated with the Strait of Gibraltar, and was ruled by a powerful and virtuous monarchy.

• Image 1: Visual Material •

Atlantis was described as a utopian society, possessing unparalleled advancements in technology, architecture, and governance for its time. The society was built around concentric rings of water and land, designed with exceptional engineering prowess and a sophisticated system of canals and irrigation to sustain its agricultural abundance.

The rulers of Atlantis were depicted as noble and enlightened, governing with wisdom and benevolence. They established a society that was highly sophisticated in terms of knowledge, culture, and military strength, often seen as a pinnacle of human achievement.

However, Plato's account also details a tragic and sudden demise of Atlantis. He narrates that approximately 9,000 years before his era, the island was struck by a catastrophic event, causing its swift destruction. The nature of this catastrophe, whether it was a natural disaster, such as an earthquake or volcanic eruption, or a mythical allegory representing moral decay, remains a topic of speculation and debate among scholars.

Plato's descriptions of Atlantis have inspired countless theories and quests to uncover its possible location. Some researchers interpret Atlantis as a purely allegorical or fictional tale, a metaphorical story used by Plato to convey philosophical or political ideas. Others have embarked on explorations seeking to find a real geographical location that matches the descriptions given by Plato, leading to various hypotheses about its potential existence in regions such as the Mediterranean, the Caribbean, or even Antarctica.

The allure of Atlantis lies in its depiction as an advanced and enigmatic civilization that met a catastrophic fate, leaving behind an enduring mystery that continues to captivate the imagination of

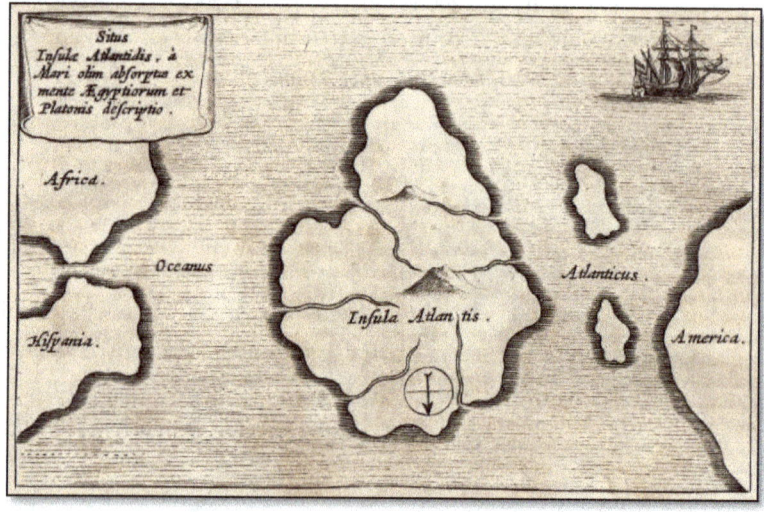

• Image 2: Athanasius Kircher's map of Atlantis, 1669 •

scholars, historians, and enthusiasts across the ages. Despite extensive research and exploration, the true nature and whereabouts of Atlantis, whether a historical reality or an allegorical invention, remain elusive, contributing to its ongoing mystique and fascination.

One of the most prominent theories proposes a connection between the Atlantis story and the volcanic eruption that devastated the island of Santorini (Thera) around 1600 BCE. This cataclysmic event resulted in the collapse of the Minoan civilization and might have inspired Plato's narrative. However, discrepancies in timelines and geographical descriptions cast doubt on this theory's direct link to Atlantis.

Additionally, the absence of tangible artifacts or ruins corresponding to the intricate layout described by Plato raises skepticism among historians and archaeologists. The lack of archaeological evidence, combined with the fantastical nature of Plato's narrative, leads many to view Atlantis as a philosophical allegory embodying Plato's ideals rather than a genuine historical account.

Scientific inquiry continually challenges the existence of Atlantis as a physical reality. While the search for Atlantis fuels the imagination and drives exploration, scientific scrutiny reminds us of the importance of critical analysis and empirical evidence in unraveling historical mysteries.

• Image 3: Visual Material •

The fascination with Atlantis extends far beyond the mere search for a lost city; it encompasses profound inquiries into the broader themes of human civilization, historical preservation, and the relationship between human achievements and the natural world.

Firstly, the story of Atlantis invokes contemplation on the rise and fall of civilizations. It serves as a reminder that even the most advanced and flourishing societies are not immune to downfall. The account of Atlantis illustrates how civilizations, no matter how prosperous or technologically advanced, can meet sudden and catastrophic ends. This narrative prompts reflection on the vulnerabilities and fragility inherent in the trajectory of human societies, reminding us that great achievements can be fleeting in the face of unforeseen or uncontrollable circumstances.

Secondly, the mystery of Atlantis raises intriguing questions about historical memory and the transmission of knowledge across generations. Plato's dialogues, "*Timaeus*" and "*Critias*," present Atlantis as a society that existed thousands of years before his time. The fact that this ancient tale has persisted through centuries, capturing the imagination of people across diverse cultures and epochs, underscores the importance of historical narratives and the enduring impact of storytelling in preserving collective memory. It highlights how stories, myths, and legends continue to shape our understanding of the past and influence our perspectives on history, even if their factual accuracy remains uncertain.

• Image 4: Visual Material •

Moreover, Atlantis sparks contemplation about the relationship between human accomplishments and the forces of nature. Whether interpreted as a cautionary tale or a mythical allegory, the catastrophic fate of Atlantis prompts reflection on humanity's relationship with the environment. It raises awareness about the potential consequences of human actions on the natural world and emphasizes the importance of balance, sustainability, and resilience in the face of environmental challenges.

The enduring allure of Atlantis lies in its ability to provoke profound reflections on the cyclical nature of civilizations, the transmission of historical knowledge, and the delicate balance between human achievements and the powerful forces of nature. Its legacy transcends the boundaries of a mere lost city, inviting contemplation and discussion about broader themes that resonate with the human experience across time and cultures.

THE ENIGMATIC NAZCA LINES

In the vast, arid deserts of southern Peru lies a puzzling testament to an ancient civilization's creativity and precision—the Nazca Lines. Etched into the desert floor over 2,000 years ago, these immense geoglyphs, depicting a wide array of figures, patterns, and shapes, have baffled researchers and explorers for centuries.

• Image 5: The Monkey •

Spread across the Nazca Desert, these lines, ranging from simple geometric shapes to elaborate designs of animals, birds, and humanoid figures, were crafted by removing the reddish-brown iron oxide-coated pebbles, revealing the lighter-colored ground beneath. The scale and precision of these formations, visible only from above, raise profound questions about the purpose and methods behind their creation.

Theories abound regarding the Nazca Lines' purpose. Some propose ceremonial or religious significance, suggesting that they served as pathways for ceremonial processions or rituals, aligning with celestial events like solstices or acting as offerings to deities. Others speculate that they might have functioned as giant astronomical calendars or served as markers for underground water sources in this arid region.

Advancements in technology have allowed researchers to uncover additional figures hidden beneath the surface, revealing a complex network of geoglyphs beyond what was initially visible. This discovery hints at a more extensive and intricate system of designs, deepening the mystery of the Nazca Lines' purpose and the civilization that created them.

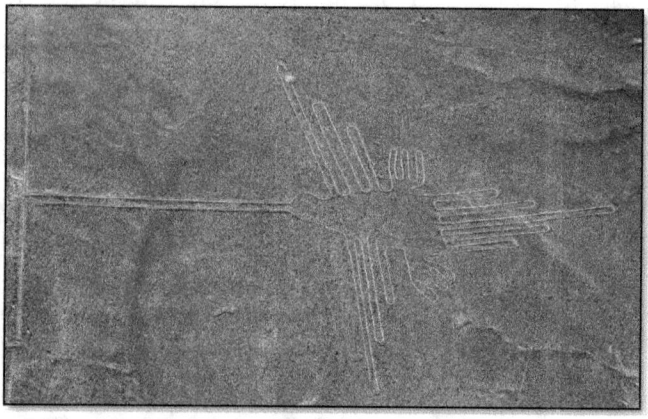

• Image 6: The Hummingbird •

Despite extensive research and study, the precise methods utilized by the ancient Nazca people to create the monumental figures remain shrouded in mystery. The sheer accuracy and scale of these enormous geoglyphs, often spanning hundreds of meters across the desert landscape, continue to baffle modern scientists and archaeologists. One of the most perplexing aspects is the remarkable precision in crafting these immense designs without the aid of aerial views or advanced technology. The Nazca people didn't possess the modern tools and equipment available to us today, yet they achieved an astonishing level of accuracy in shaping these intricate and vast figures.

Scholars have proposed several theories about how the Nazca lines were created. Some suggest that the Nazca might have used simple tools like ropes, stakes, and basic surveying techniques to create the outlines of the figures. Others speculate that the Nazca might have employed mathematical and astronomical knowledge, aligning their designs with celestial bodies or using natural landmarks to guide their construction.

The process of creating the Nazca lines involved removing the dark surface stones to reveal the lighter-colored soil beneath, resulting in the stark contrast that defines these figures. The most prominent Nazca geoglyphs depict animals, birds, and geometric shapes, showcasing a diverse range of intricate designs executed with remarkable precision and symmetry.

• Image 7: The Spider •

Furthermore, the sheer scale and complexity of these geoglyphs suggest meticulous planning and organization by the Nazca people. The accuracy of these designs, often with straight lines and perfectly shaped curves, challenges our understanding of ancient engineering and artistic capabilities.

The enigma of the Nazca lines continues to stimulate scholarly debate and inspires researchers to delve deeper into understanding the methods and motivations behind their creation. The Nazca geoglyphs stand as a testament to the ingenuity and creativity of ancient civilizations, leaving behind a legacy that intrigues and captivates our imagination while posing enduring questions about the technological and artistic prowess of our distant ancestors.

The enduring mystery surrounding the Nazca Lines lies in the ambiguity of their purpose and the intricate societal motivations that led to their creation. Despite numerous hypotheses and studies, the true intent behind these colossal designs continues to elude definitive explanation.

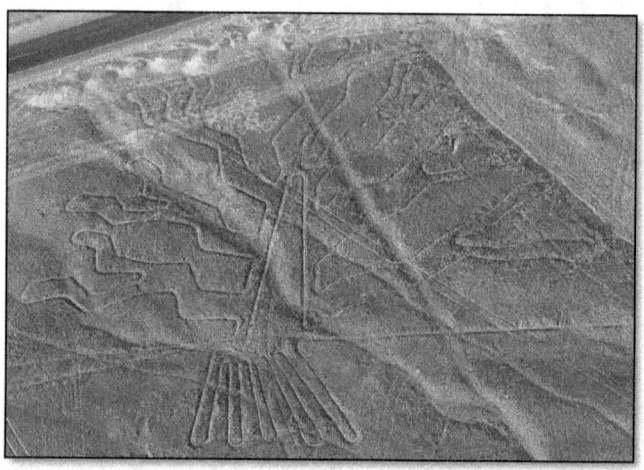

• Image 8: The Tree •

The multifaceted nature of the Nazca Lines presents a complex puzzle that intertwines art, engineering, spirituality, and cultural expression. One prevalent theory suggests that these geoglyphs might have held ritualistic or ceremonial significance for the Nazca people. It's speculated that these figures could have been linked to their spiritual beliefs, serving as pathways for ceremonial processions or as offerings to deities, aligning with the natural landscape and celestial events.

Another school of thought proposes that the Nazca Lines might have served as an astronomical calendar or a means to track celestial movements, aligning with the solstices or equinoxes. The precision and orientation of some geoglyphs in relation to astronomical events lend credibility to this theory, suggesting a possible connection between the Nazca people's religious or agricultural calendar and these enigmatic designs.

Furthermore, these colossal figures might have symbolized the Nazca people's reverence for nature or their surroundings, representing animals, plants, and geometric shapes that held cultural or mythological significance. Some interpretations suggest that the geoglyphs conveyed messages about the environment, water sources, migration routes, or marked sacred places within their belief system.

The Nazca Lines, with their diverse motifs and grand scale, continue to ignite the imagination, prompting us to explore the complexities of ancient cultures and their connections with the natural and spiritual realms. These enigmatic designs serve as a gateway for us to peer into the depths of history, inviting contemplation on the rich tapestry of human ingenuity, belief systems, and artistic expression that flourished in ancient times.

Ultimately, the true purpose and meaning behind the Nazca Lines may forever remain an enigma, allowing room for multiple interpretations and encouraging ongoing exploration and discourse among researchers and enthusiasts alike. Their enduring legacy serves as a testament to the depth of human creativity and the profound mysteries that continue to intrigue and inspire us across generations.

The Nazca Lines continue to captivate, raising more questions than answers and serving as a testament to the ingenuity and creativity of ancient civilizations. As we unravel the mysteries of these enigmatic geoglyphs, we delve deeper into the intricate tapestry of human history, contemplating the remarkable achievements of those who came before us.

The Roanoke Colony stands as a poignant tale, shrouded in mystery and leaving an indelible mark on the landscape of early American colonization. The story begins in 1587 when John White led a group of English settlers to establish a foothold in the New World, aiming to create a prosperous colony in what is now North Carolina.

The trajectory of the Roanoke Colony took an unforeseen and dramatic turn when John White, the governor of the settlement, confronted a critical situation characterized by depleting supplies and the urgent necessity for external aid. Faced with this dire circumstance, White made the difficult decision to sail back to England in 1587, seeking assistance and additional resources to sustain the struggling colony.

● Image 9: «John White at the ruins of the
Roanoke colony», 1590 ●

However, upon his arrival in England, unforeseen and challenging circumstances unfolded. England was embroiled in a conflict with Spain, which significantly impacted the nation's resources and priorities. The ongoing war diverted attention and resources away from supporting overseas colonies, presenting formidable obstacles to White's efforts in securing immediate aid for the beleaguered settlers in Roanoke.

Furthermore, logistical challenges and difficulties in acquiring suitable vessels, supplies, and the necessary provisions for the colony compounded the delays in John White's return. The process of organizing and assembling a relief expedition to assist the Roanoke Colony was fraught with obstacles, causing significant delays that extended White's absence far beyond what was initially anticipated.

As the months turned into years, the colonists in Roanoke faced mounting challenges and uncertainties in the absence of crucial supplies and reinforcements. The prolonged delay in John White's return exacerbated the precarious situation of the settlement, leaving the fate of the colonists hanging in uncertainty and vulnerability to the harsh realities of survival in an unfamiliar and challenging environment.

The unexpected three-year hiatus in White's return to Roanoke became a pivotal moment in the colony's history, marking a period of prolonged isolation and deprivation that undoubtedly influenced the subsequent events and the ultimate mysterious disappearance of the Roanoke settlers. The combination of White's extended absence, England's wartime circumstances, and logistical hurdles formed a crucial backdrop to the colony's enigmatic fate, leaving an indelible mark on the historical narrative of the lost Roanoke Colony.

• Image 10: Depicting the discovery of the abandoned colony •

Upon his long-awaited return in 1590, John White was met with an eerie silence. The settlement was eerily deserted, devoid of its inhabitants. The only semblance of a clue was the word "*Croatoan*" carved into a wooden post, leaving behind an enigma that has perplexed historians and researchers for centuries.

The disappearance of the Roanoke Colony has sparked a myriad of theories attempting to unlock its secrets. Among them is the hypothesis that the settlers might have sought refuge or integrated with nearby Native American tribes, possibly the Croatan people, suggested by the inscribed word. Others propose tragic endings, such as conflicts with indigenous communities, famine, disease, or internal disputes among the settlers themselves.

Despite relentless efforts through extensive investigations and continued archaeological excavations, the quest to unravel the fate of the Roanoke settlers has encountered an enduring impasse. The search for concrete evidence shedding light on the ultimate destiny of these early American colonists has proved to be an elusive endeavor.

Archaeologists, historians, and researchers have dedicated substantial resources and expertise to unraveling the enigma of the Roanoke Colony. Excavations conducted at the site and its vicinity have yielded valuable insights and uncovered artifacts shedding some light on the daily lives of the settlers. However, the crucial evidence required to definitively explain their fate remains conspicuously absent.

The absence of conclusive artifacts, such as personal belongings, tools, or structures distinctly linked to the missing colonists, complicates the efforts to piece together their story. The lack of documented accounts or verifiable narratives from that era further deepens the mystery, leaving historical narratives with significant gaps and unanswered questions regarding their sudden and complete disappearance.

Various theories and conjectures have emerged over time in attempts to explain the fate of the Roanoke settlers. Some hypothesize that the colonists might have integrated or merged with local Native American tribes for survival, while others suggest the possibility of clashes with indigenous communities or hardships leading to their dispersion or migration to remote areas.

However, without tangible evidence to substantiate these theories, the mystery persists, perpetuating the enduring fascination with the Roanoke Colony's vanishing. The absence of definitive proof has led to a proliferation of speculation and a multitude of hypotheses, each offering a fragmentary perspective but failing to provide a conclusive resolution to the haunting puzzle of the vanished colony.

● Image 11: Colonists arrive in America ●

The elusive nature of the evidence surrounding the Roanoke settlers' disappearance continues to stoke the imagination of historians, sparking renewed interest and ongoing efforts to uncover new clues or perspectives that might one-day illuminate the fate of these early American pioneers. Until concrete evidence emerges, the enigma of the Roanoke Colony's mysterious vanishing remains entrenched in the annals of history, reminding us of the profound mysteries that history sometimes chooses to withhold.

The Roanoke Colony's story embodies the complexities and uncertainties of early colonial ventures, serving as a somber reminder of the trials faced by those who sought to establish footholds in unknown territories. The unanswered questions surrounding the fate of the settlers echo through history, inviting contemplation and stirring the imagination of those who ponder the past.

As with many historical enigmas, the Roanoke Colony disappearance raises profound questions about the interplay between cultures, the challenges of early colonization, and the intrinsic human drive for exploration and settlement.

• Image 12: Map of Roanoke Island and surrounding lands •

The tantalizing puzzle of Roanoke's vanishing continues to provoke scholarly debate and capture the imagination of enthusiasts, serving as a testament to the enduring allure of unsolved historical mysteries. Each attempt to decipher its riddle offers a glimpse into the complexities of human history and the inherent uncertainties that accompany our quest to understand the past.

The Roanoke Colony's disappearance stands as a testament to the limits of our historical knowledge, reminding us that amidst the passage of time, some mysteries may persist, eternally challenging our understanding of the past and the enigmatic tales that shape our collective narrative.

In our ongoing exploration of historical enigmas, we invite readers to join us in unraveling the threads of lost civilizations, delving into cryptic creatures, pondering ancient puzzles, and contemplating the enigmatic phenomena that have fascinated and perplexed humanity throughout the ages.

As we embark on this journey through history's mysteries, let us embrace the quest for understanding while acknowledging the enigmatic allure of the unknown that continues to spark our curiosity and drive our exploration of the past.

UNVEILING THE OLMEC'S - AN ANCIENT ENIGMA

Within the pages of Mesoamerican history, the Olmec's emerge as enigmatic pioneers of an ancient civilization, nestled amidst the lush landscapes of modern-day Mexico. Regarded as trailblazers of sophisticated societies in the area, their enduring legacy echoes through the ages, veiled in enigmatic mysteries that enshroud their culture and impact.

The Olmec civilization thrived from around 1500 BCE to 400 BCE, situated in the tropical lowlands of the Gulf of Mexico, particularly in what is now the Mexican states of Veracruz and Tabasco. They constructed grand ceremonial centers, notably la Venta, San Lorenzo, and Tres Zapotes, indicative of their sophisticated architectural prowess.

• Image 13: The colossal stone head is 17 feet tall •

Central to Olmec iconography are the colossal stone heads, sculpted from volcanic rock, showcasing distinctive facial features and headdresses, each weighing several tons. These monumental sculptures, discovered in various regions, continue to baffle historians, inspiring debates regarding their purpose and the civilization's societal hierarchy.

The Olmec's agricultural expertise was a cornerstone of their civilization's success. They ingeniously crafted elaborate irrigation and drainage systems, showcasing their advanced knowledge of land cultivation. These sophisticated systems allowed them to tame and cultivate the fertile lands along the Gulf of Mexico, enabling efficient farming of staple crops like maize (corn), beans, and squash. The mastery of agriculture facilitated surplus food production, sustaining their population and fostering social development.

Moreover, the Olmec's proficiency in trade played a pivotal role in their society. They engaged in extensive commercial activities, exchanging commodities both within their community and with neighboring regions. Among their traded goods were precious materials like jade, known for its symbolic and spiritual significance, obsidian utilized for crafting tools and weaponry, and rubber, obtained from the latex of local trees, which held multiple practical applications.

This intricate network of trade routes and economic exchanges underscored the Olmec's influence in regional commerce, allowing them to establish relationships with neighboring cultures and cementing their significance in the broader Mesoamerican trade network. The flow of goods and ideas enriched their culture and further enhanced their societal development.

• Image 14: «The Wrestler» statuette, dating back to 1400–400 BCE •

Archaeological excavations have revealed the Olmec's complex social structure and religious beliefs. Their artwork, pottery, and carvings depict ritualistic ceremonies, zoomorphic deities, and enigmatic symbols that suggest a sophisticated cosmology.

Scholars hypothesize that the Olmec's may have influenced subsequent Mesoamerican civilizations, such as the Maya and Aztecs, through their architectural styles, religious ideologies, and trade networks.

Scientific analysis of Olmec artifacts and monuments, including hieroglyphic inscriptions, has posed challenges in deciphering their written language. The lack of a comprehensive understanding of Olmec script remains a significant obstacle in unraveling their intellectual and cultural heritage.

● Image 15: "Parque Museo" la Venta ●

The Olmec's contribution to Mesoamerican civilization is undeniable, yet much about their society remains shrouded in ambiguity. Questions surrounding their sudden decline around 400 BCE and the factors contributing to their demise persist, spurring ongoing archaeological investigations and academic discourse.

In summary, the Olmec's, with their enigmatic stone monuments, sophisticated culture, and influential legacy, continue to fascinate researchers and enthusiasts alike. Their story represents a crucial piece in the intricate tapestry of ancient civilizations, urging us to probe deeper into the mysteries that veil their remarkable past.

As we traverse the fog of time and delve into their remarkable saga, let us acknowledge the Olmec's profound impact on Mesoamerican history while continuing our quest to unveil the secrets of this lost civilization.

CHAPTER 2: CRYPTIC CREATURES

Within the domain of enigmatic beings, we embark on an exploration of elusive creatures like Bigfoot, dwelling deep within forests; the mysterious Loch Ness Monster, said to inhabit the depths of Scottish waters; and the eerie Chupacabra, a cryptic figure rooted in folklore. These elusive entities spark wonder and legend, blurring the boundaries between fact and fantasy. Join us as we venture into the captivating realm of cryptic creatures, investigating the myths, sightings, and enduring allure encircling these enigmatic entities.

"The allure of cryptic creatures lies in their ability to stir our curiosity, challenging our perceptions of the natural world"

Cryptozoologist Loren Coleman

BIGFOOT: MYTH OR REALITY?

The legend of Bigfoot, or Sasquatch, has entrenched itself as one of the most enduring and contentious mysteries in the domain of cryptozoology. This enigmatic creature's presence transcends geographical boundaries, woven into the tales and narratives of diverse cultures and regions worldwide. Stories abound of a colossal, ape-like being traversing dense forests, remote mountain ranges, and untamed wilderness, captivating the collective imagination of both ardent believers and skeptical observers.

Reports and accounts of encounters with Bigfoot, also known as Sasquatch, extend back across centuries, interwoven within the oral traditions and folklore of numerous indigenous cultures across different regions. Indigenous communities from North America, including the Coast Salish, Skwxwú7mesh, and Sts'ailes tribes, among others, have passed down stories of a large, hairy, and elusive humanoid creature inhabiting the wilderness, often characterized as a guardian of the forests or a figure within their cultural narratives.

• Image 16: Frame of the 1967 •

In the latter part of the 20th century, the fascination with Bigfoot garnered increased attention, largely fueled by purported sightings, footprint discoveries, and an abundance of anecdotal evidence. The surge in reported encounters and interest in the phenomenon led to an intensified quest for scientific validation and understanding.

One significant case contributing to the contemporary fascination with Bigfoot occurred in Bluff Creek, California, in 1967. This event is associated with the Patterson-Gimlin film, a short video footage shot by Roger Patterson and Bob Gimlin purportedly capturing a bipedal, ape-like creature striding along Bluff Creek. Though subject to skepticism and debate, this footage remains one of the most iconic and hotly debated pieces of evidence within the realm of Bigfoot research.

Numerous eyewitness accounts also contribute to the Bigfoot lore. For instance, in 1924, the story of Fred Beck and his mining team in Ape Canyon, Washington, gained attention. The group claimed to have encountered a group of apelike creatures, which they described as throwing rocks at their cabin during a night of terrifying encounters. Additionally, the 1974 case of the "Fouke Monster" in Fouke, Arkansas, brought the creature further into the public eye, with multiple sightings and reports from local residents describing encounters with a large, hairy creature.

• Image 17: Supposed trail Bigfoot •

Various Native American communities have shared their oral traditions and experiences with Sasquatch, describing encounters and cultural beliefs tied to these elusive beings. These stories often highlight the creature's ability to move stealthily through the forests, emphasizing its role as a guardian of nature or a spiritual entity intertwined with the natural world.

The era witnessed a surge in investigations, both amateur and professional, into the phenomenon of Bigfoot. Expeditions were organized, seeking to uncover tangible evidence that might validate the existence of this elusive creature. Researchers, cryptozoologists, and enthusiasts immersed themselves in analyzing purported footprints, collecting hair samples, and documenting eyewitness testimonies in pursuit of conclusive proof.

Nevertheless, the quest to firmly establish the reality of Bigfoot encountered significant challenges. Despite the abundance of anecdotal reports and purported evidence, scientific validation of the existence of such a creature remained elusive. Skeptics contended that the evidence presented often lacked verifiability, attributing sightings to misidentifications of known wildlife, hoaxes, or the embellishments of folklore.

The defining characteristic of Bigfoot lore is the consistent description of a creature standing between 6 to 9 feet tall, covered in dark or reddish-brown fur, possessing remarkable physical strength, and emitting a pungent odor. Witnesses often describe its upright, bipedal gait, human-like features, and a piercing, almost haunting, howl echoing through the forested landscapes.

The debate surrounding the existence of Bigfoot is deeply polarized. Advocates of its existence point to a myriad of sightings, footprint casts, and blurry photographs or videos as evidence supporting the creature's existence. They argue that vast, remote wilderness areas still remain unexplored, providing potential habitats for such elusive beings.

Conversely, skeptics argue that the lack of irrefutable physical evidence, such as a specimen, bones, or DNA samples, undermines the credibility of the Bigfoot phenomenon. They attribute alleged sightings to misidentifications of known wildlife, hoaxes, or the product of folklore and the human inclination towards myth and legend.

Scientific scrutiny of the Bigfoot legend has indeed been constrained by the absence of concrete and irrefutable evidence. Despite this hurdle, a committed community of researchers, cryptozoologists, and devoted enthusiasts persist in their unwavering pursuit of substantiating the existence of Bigfoot.

The primary challenge faced by scientific investigation into the Bigfoot phenomenon is the dearth of empirical evidence. The lack of a specimen, fossil records, skeletal remains, or biological samples that can be conclusively attributed to Bigfoot hampers rigorous scientific examination. This absence of tangible evidence presents a significant hurdle in the scientific community's efforts to validate the existence of this elusive creature.

• Image 18: Visual Material •

However, this limitation has not deterred passionate researchers and enthusiasts from actively engaging in the quest to unravel the mystery of Bigfoot. Despite the skepticism prevalent in mainstream science, a dedicated cadre of individuals continues to conduct field research, employing diverse scientific methodologies and cutting-edge technologies in their pursuit.

One approach involves employing advanced imaging technology, such as drones equipped with high-resolution cameras, LiDAR (Light Detection and Ranging), and thermal imaging devices. These tools aid researchers in conducting aerial surveys and ground-level searches of remote and densely forested areas where alleged Bigfoot sightings have been reported. Furthermore, the use of trail cameras, which automatically capture images upon detecting motion, has been deployed in the hopes of obtaining photographic or video evidence.

Cryptozoologists and researchers also engage in footprint analysis, examining purported Bigfoot footprints for anatomical details, dermal ridges, and distinctive characteristics that might distinguish them from known animal tracks. Additionally, DNA analysis of collected hair samples and biological materials purportedly linked to Bigfoot are subject to rigorous scientific scrutiny in specialized laboratories.

While the scientific community remains cautious about investing substantial resources into investigating the Bigfoot phenomenon due to the lack of credible evidence, these dedicated individuals persist in their pursuit. Their efforts reflect an ongoing commitment to scientific rigor and an unwavering determination to unravel the mystery behind one of the most enduring cryptids, Bigfoot, employing scientific methodologies and advancements in the quest for substantiation.

THE ENIGMATIC LEGEND OF THE LOCH NESS MONSTER

The legend of the Loch Ness Monster, often affectionately referred to as "Nessie," has captured the imagination and curiosity of people worldwide for generations. Nestled amidst the breathtaking landscapes of the Scottish Highlands, Loch Ness stands as the stage for one of the most enduring and debated mysteries in the realm of cryptids.

The enduring tales of a mysterious entity inhabiting the murky depths of Loch Ness have woven an intricate tapestry of folklore and intrigue across the ages. These narratives, deeply rooted in the annals of Scottish oral traditions and ancient folklore, resonate with accounts that stretch back through the corridors of time, sketching the silhouette of a creature that has etched itself into the fabric of Highland legend.

Centuries-old stories speak of encounters with a creature resembling a serpentine being or an entity reminiscent of a prehistoric dinosaur. The lore paints vivid pictures of an enigmatic inhabitant prowling the fathomless waters of Loch Ness, occasionally breaking the surface in fleeting moments, leaving startled witnesses awestruck by glimpses of its mysterious form.

• Image 19: Loch Ness Monster in 1934 •

The genesis of the Loch Ness Monster legend lies entwined with the rich tapestry of Scottish mythology, where tales of fantastical beings and mystical creatures find their place among the traditions passed down through generations. Ancient folklore, woven intricately with the essence of the Highlands, speaks of an otherworldly dweller inhabiting the depths of the fabled Loch Ness.

Descriptive accounts vary, but a common thread weaves through these tales: the portrayal of a creature exuding an air of mystery and intrigue. Some describe a creature bearing the semblance of a serpent, with a sinuous form gliding stealthily through the depths. Others evoke imagery of a large, reptilian being reminiscent of prehistoric dinosaurs, fueling the imagination and curiosity of those who hear these tales.

However, it was in the early 20th century that the legend of the Loch Ness Monster garnered international attention. In 1933, an intriguing photograph captured by Hugh Gray depicting what appeared to be a long-necked creature emerging from the water catapulted Nessie into the global spotlight, sparking widespread curiosity and speculation. This iconic image fueled a surge of interest, and numerous purported sightings and accounts of encounters followed suit, cementing the Loch Ness Monster as a fixture in popular culture.

The descriptions of the Loch Ness Monster vary among eyewitness testimonies, with some accounts portraying it as a large aquatic creature with a humped back and a long neck, akin to a plesiosaur or an unknown marine reptile from prehistoric times. Others describe it as a serpentine beast or a sizable creature with a mysterious and elusive presence, adding layers of mystique to the legend.

• Image 20: Visual Material •

The fascination with Nessie has prompted extensive investigations, scientific inquiries, and numerous expeditions aimed at unraveling the mystery behind this elusive creature. Researchers, cryptozoologists, and enthusiastic amateurs have employed various technological advancements, including sonar imaging, underwater cameras, and sophisticated surveillance equipment, to probe the depths of Loch Ness in search of tangible evidence.

The quest for irrefutable evidence substantiating the existence of the Loch Ness Monster has spanned decades, marked by persistent efforts and advancements in scientific methodologies. However, despite these endeavors, a definitive and conclusive proof validating the existence of Nessie has continued to evade researchers and enthusiasts alike.

Skeptics consistently offer alternative explanations for reported sightings, sighting a range of possibilities. Misidentifications of known animals, such as otters, large fish, or birds, often feature prominently among these explanations. Additionally, floating debris or natural phenomena, coupled with the interplay of light and shadows on the lake's surface, are frequently cited as potential causes for the perceived sightings of a mysterious creature.

• Image 21: Visual Material •

The absence of tangible and unequivocal empirical evidence has spurred vigorous debates within the scientific community and the public sphere, leading to a pronounced divide between proponents of Nessie's existence and staunch skeptics. Believers in the Loch Ness Monster's reality often point to anecdotal accounts, eyewitness testimonies, and occasional photographs or videos as compelling albeit circumstantial evidence supporting the cryptid's presence in the Scottish loch. However, skeptics challenge the validity of these accounts, emphasizing the subjective nature of eyewitness observations and the lack of scientifically verifiable proof.

Scientific endeavors to unveil the truth behind Nessie's existence have encompassed a wide array of approaches. Sonar technology, deployed in extensive underwater surveys, has been utilized to scan the depths of Loch Ness in search of large unidentified objects or unusual anomalies. High-resolution underwater cameras and surveillance equipment have also been employed in attempts to capture definitive visual evidence of the alleged creature. However, these efforts, while yielding intriguing data and occasional intriguing sonar readings, have not provided the conclusive evidence needed to definitively confirm the Loch Ness Monster's existence.

● Image 22: Visual Material ●

As we embark on an exploration into the enigmatic world of cryptids, the legend of the Loch Ness Monster emerges as an enduring mystery that beckons us to delve deeper into the realms of folklore, scientific inquiry, and the age-old fascination with elusive creatures that continue to intrigue and captivate our imagination.

CHUPACABRA: ELUSIVE PREDATOR OF FOLKLORE

In the realm of cryptids, few creatures have ignited as much intrigue, fear, and speculation as the legendary Chupacabra. Emerging from the vast tapestry of folklore and whispered tales, this enigmatic predator has woven itself into the fabric of cultural narratives across various regions, evoking curiosity and fascination mixed with a tinge of trepidation.

The legend of the Chupacabra finds its origins steeped in the folklore of Latin American countries, particularly Puerto Rico, where reports of peculiar livestock attacks surfaced in the latter part of the 20th century. The term "Chupacabra" translates to "goat-sucker" in Spanish, a name reflective of its alleged vampiric tendencies of preying on livestock, particularly goats and other small animals, and supposedly draining them of blood.

Accounts depicting encounters with the Chupacabra paint a picture of a creature enshrouded in enigmatic details, fostering a diverse spectrum of descriptions that have fueled the mystique surrounding this elusive predator. The reported physical characteristics of the Chupacabra vary significantly across different sightings and cultural narratives, contributing to the perplexity and intrigue surrounding its alleged existence.

Some accounts convey a portrayal of the Chupacabra as a reptilian entity, evoking images of scaled skin, a sleek and serpentine body, and a snakelike or lizard-like appearance. Others describe the creature with canine attributes, resembling a fearsome hybrid between a dog or wolf and a creature with distinctly unearthly features. Such depictions often feature a bipedal stance, with the Chupacabra purportedly standing on two legs, exhibiting an eerie humanoid posture that defies conventional zoological classification.

• Image 23: Visual Material •

The recurring motif in these accounts is the portrayal of large, piercing, and luminescent red eyes that seem to glow in the darkness—a detail that instills a sense of dread and otherworldly quality to the Chupacabra's visage. Witnesses often describe the creature's gaze as hypnotic or unnerving, leaving an indelible impression on those who claim to have encountered it.

Descriptive narratives also commonly attribute the Chupacabra with formidable features such as sharp, elongated fangs or teeth, resembling those of a predator capable of inflicting lethal injuries on its prey. Some reports mention peculiar spikes, quills, or spines running along the creature's back, adding an element of the bizarre and alien to its appearance—a feature that has stirred comparisons to extraterrestrial beings in the annals of cryptozoological lore.

However, despite the array of descriptions that vary in physical attributes, a consistent thread interweaves these accounts—

the element of fear and fascination evoked by the mere mention of the Chupacabra

The emergence of the Chupacabra legend traces its origins to a series of puzzling incidents that surfaced in the mid-1990s, primarily in regions of Latin America and particularly Puerto Rico. These incidents revolved around a perplexing phenomenon marked by reports of unusual and unexplained animal mutilations, coupled with the peculiar circumstance of discovering carcasses drained of blood.

• Image 24: Visual Material •

Initially, the reports centered on instances of local farmers finding their livestock—predominantly goats, sheep, and other small animals—mysteriously killed and displaying distinctive injuries. The hallmark of these incidents was the absence of traditional predator marks or traces of typical scavenger activity, confounding both local residents and authorities alike.

What set these occurrences apart was the absence of blood in the animal carcasses, adding an eerie and perplexing element to the situation. These bloodless carcasses became a defining characteristic of the purported Chupacabra attacks, elevating the mysterious nature of these events and intensifying public intrigue and concern.

The impact of these accounts was swift and profound. The media, both local and international, amplified these reports, disseminating them through news outlets, tabloids, and word of mouth. The sensational nature of the Chupacabra narrative captured the public's imagination, resonating deeply within the cultural consciousness and swiftly permeating popular culture.

• Image 25: Visual Material •

In the wake of these reports, fervent investigations and speculative theories emerged, attempting to unravel the mystery behind the Chupacabra's existence. Cryptozoologists, researchers, and curious enthusiasts embarked on a quest to decipher the truth behind this elusive creature. However, the Chupacabra, with its elusive nature and disparate descriptions, continues to evade conclusive identification or scientific verification, relegating it to the sphere of cryptids—mysterious beings that exist at the edge of belief and skepticism.

The legend of the Chupacabra, steeped in folklore and cultural mythos, stands as a testament to the enduring allure of cryptids that permeate cultural narratives and spark the human imagination. As we delve deeper into the mysteries of cryptic creatures, the legend of the Chupacabra emerges as a compelling enigma, beckoning us to explore the blurred boundaries between folklore, reality, and the realms of the unknown.

CHAPTER 3: ASTOUNDING ARCHAEOLOGICAL DISCOVERIES

Step into the world of astonishing archaeological discoveries, where we unravel the mysteries behind captivating finds like the enigmatic Stonehenge, the unresolved puzzles encompassing the Pyramids, and the monumental Moai statues of Easter Island. These intriguing revelations speak volumes about the creativity and unanswered questions of ancient civilizations, inviting us to ponder their significance. Join us on a captivating expedition into the realm of remarkable archaeological wonders, where each artifact holds a piece of humanity's intriguing history.

"Astounding archaeological finds are the whispers of antiquity, speaking volumes about civilizations lost to time"

Zahi Hawass

UNRAVELING THE ENIGMA OF STONEHENGE

Stonehenge, the ancient monument that stands as a testament to prehistoric ingenuity and human endeavor, has captivated the imagination of scholars, archaeologists, and curious minds for centuries. Nestled upon the vast Salisbury Plain in England, this enigmatic structure embodies a mystery that spans millennia, evoking wonder and contemplation about its purpose, construction, and the civilization that gave birth to its existence.

• Image 26: Original photo •

Stonehenge's genesis traces back approximately 4,000 years to the late Neolithic period and the Bronze Age, marking the commencement of its construction. This monumental site, situated on the Salisbury Plain in Wiltshire, England, stands as an extraordinary feat of ancient architecture and engineering, capturing the imagination with its enigmatic design and colossal stone structures.

At its core, Stonehenge comprises an arrangement of massive standing stones, consisting of two primary stone types: the larger sarsens, weighing up to 25 tons and standing as high as 30 feet, and the smaller bluestones, transported from the Preseli Hills in Wales, each weighing several tons. The arrangement of these stones forms a circular and concentric layout, accompanied by lintels, creating the iconic structure that continues to fascinate and mystify.

Scientists and researchers have drawn several conclusions regarding Stonehenge, although many aspects of its construction and purpose continue to intrigue and elude complete understanding. Some key conclusions drawn by scientists based on archaeological studies, analyses, and ongoing research include:

- *Ceremonial and Ritualistic Site:* Stonehenge was likely a place of significant ceremonial and ritualistic importance for the prehistoric people of Britain. Its alignment with celestial events, such as the solstices, suggests a link to seasonal rituals or astronomical observations.

- *Astronomical Significance:* The precise alignment of Stonehenge's stones and features with celestial phenomena, particularly the summer and winter solstices, has led scientists to believe that it might have functioned as an astronomical observatory, tracking celestial movements or marking important solar and lunar events.

- *Cultural Significance:* Stonehenge reflects the architectural and engineering capabilities of its builders. The transportation and arrangement of massive stones, some weighing several tons, signify the technological prowess and communal efforts of the Neolithic and Bronze Age people who constructed it.

- *Funerary and Burial Practices:* Excavations around Stonehenge have revealed burial mounds and evidence of cremations, indicating that it might have served as a burial site or been associated with rituals related to the deceased.

- *Continued Use and Evolution:* Evidence suggests that Stonehenge's significance evolved over time, with multiple construction phases spanning centuries. The site might have undergone alterations, with new stones added or repositioned, indicating that its use and significance may have shifted across generations.

- *Social and Spiritual Hub:* Stonehenge likely served as a social and spiritual hub, attracting people from different regions, fostering cultural exchange, religious practices, and possibly acting as a place for healing or pilgrimage.

- *Symbol of Authority or Power:* Some theories suggest that Stonehenge might have been a symbol of authority or power, representing the might and influence of certain groups or individuals within the society of the time.

One of the prevailing theories suggests that the sarsen stones, sourced locally from Marlborough Downs, were transported overland to the site. Hypotheses propose the use of wooden sledges, rollers, and possibly even wooden structures or earthen embankments as rudimentary forms of transport to maneuver these colossal stones across vast distances.

• Image 27: Original photo •

The smaller bluestones, originating from the Preseli Hills in Wales, present an even more intriguing challenge. The geological sourcing of these stones, approximately 150 miles away from the Stonehenge site, has fueled discussions about the methods utilized to transport them to their final destination. The prevailing theory involves the use of water transport along rivers and possibly over the sea, suggesting that ancient civilizations possessed navigational skills and technologies enabling them to move these stones via waterways.

The precise techniques employed to lift and position these monumental stones remain subjects of speculation. Some hypotheses propose the use of wooden or stone tools, ropes, and lever systems, along with the collaboration of numerous laborers, to hoist these massive stones into their upright positions.

The construction of Stonehenge required not only remarkable engineering prowess but also a concerted, labor-intensive effort over an extended period, possibly spanning centuries. The precision and alignment of the stones, particularly concerning their astronomical orientations, hint at the advanced knowledge and understanding of celestial phenomena possessed by the ancient builders.

The significance of Stonehenge extends beyond its physical structure. It served as a celestial observatory, aligned with astronomical events such as solstices and equinoxes. The precision in its orientation to celestial phenomena hints at an advanced understanding of celestial mechanics by the builders, lending an air of sophistication to their civilization.

Throughout history, Stonehenge has endured, bearing witness to the passage of time and the evolution of human societies. Its purpose, however, remains shrouded in mystery, a puzzle that continues to

● Image 28: Visual Material ●

intrigue and elude definitive explanation. Theories abound about its intended function—a ceremonial site, a place of worship, an astronomical calendar, or even a burial ground—each offering a fragmentary glimpse into its possible significance.

Archaeological excavations and continuous advancements in scientific methodologies have embarked upon an intricate journey into the heart of Stonehenge, unraveling layers of its mysteries and shedding light on the lives and practices of the ancient people who constructed this iconic monument.

The meticulous examination of Stonehenge and its surrounding landscape has yielded a trove of compelling archaeological discoveries. Excavations within and around the monument have unearthed a myriad of artifacts, offering glimpses into the rituals, customs, and cultural practices of its builders. These discoveries include items such as finely crafted tools, pottery fragments, ceremonial objects, and remnants of offerings, each contributing to a deeper understanding of the society that thrived during the monument's construction.

Burial sites and human remains found in proximity to Stonehenge have provided invaluable insights into the individuals who lived in its vicinity. The examination of these burial grounds has revealed information about social structures, kinship ties, and possibly the roles of individuals within their community. Analysis of skeletal remains and grave goods has provided clues about diet, health, and possibly even aspects of belief systems prevalent during that era.

• Image 29: Original photo •

Additionally, evidence of ceremonial activities has been uncovered, offering glimpses into the spiritual and religious practices associated with Stonehenge. Ritual deposits, such as animal bones, ceremonial objects, and offerings, suggest that the monument played a central role in religious ceremonies, communal gatherings, and possibly astronomical observations linked to the cycles of the sun and moon.

Scientific techniques, including radiocarbon dating, advanced imaging technologies, and geophysical surveys, have further enhanced our understanding of Stonehenge's chronology, construction phases, and the landscape's significance in relation to the monument. These methods have allowed archaeologists to refine their interpretations and reconstruct the timeline of Stonehenge's development, enabling a more nuanced understanding of its purpose and cultural significance.

As we embark on a journey to unravel the enigma of Stonehenge, we delve into a captivating narrative woven from archaeological discoveries, scientific inquiry, and the enduring fascination with an ancient wonder that continues to spark awe and curiosity in the hearts and minds of people across the globe. Stonehenge stands not merely as a physical relic but as an enduring testament to the indomitable spirit of human curiosity and the mysteries enshrined in our distant past.

THE PYRAMID ENIGMA UNVEILED?

The Pyramids of Egypt, standing as magnificent architectural marvels on the Giza Plateau, represent an enduring enigma that has captivated humanity for millennia. These colossal structures, rising majestically from the desert sands, bear witness to the ancient Egyptians' ingenuity, engineering prowess, and their profound reverence for the afterlife.

• Image 30: Original photo •

The genesis of Egypt's pyramid construction dates back over 4,500 years to the Old Kingdom period, marking a significant era in human history. The Pyramids of Giza, including the Great Pyramid of Khufu (Cheops), the Pyramid of Khafre (Chephren), and the Pyramid of Menkaure, constitute an awe-inspiring testament to the ancient Egyptians' remarkable architectural achievements.

The Great Pyramid of Khufu, also known as the Pyramid of Cheops, remains a marvel of ancient engineering that continues to baffle researchers and enthusiasts alike. Its sheer size, precision construction, and complex design have intrigued scholars for centuries, prompting ongoing investigations into the mysteries that enshroud its creation.

Constructed during the Fourth Dynasty of Egypt around 2580–2560 BCE, the Great Pyramid stands as the largest of the three pyramids at Giza, a testament to the architectural prowess of ancient Egypt. Comprising an estimated 2.3 million limestone blocks, some weighing up to several tons each, this colossal structure is a remarkable feat of precision engineering.

• Image 31: Original photo •

The precise techniques utilized to quarry, transport, and position these immense blocks are subjects of intense scholarly debate and speculation. Ancient Egyptians likely quarried the limestone blocks from nearby quarries, possibly using copper tools and wooden sledges to transport them to the pyramid's construction site. The exact method of lifting and placing these massive stones into position remains an enigma, with theories suggesting the use of ramps, levers, and counterweights or possibly even a combination of techniques employed by skilled laborers.

The purpose behind the construction of the Great Pyramid has long been a topic of scholarly discussion. While traditional belief holds that these pyramids served as tombs for the pharaohs, particularly Khufu (Cheops) himself, providing a grand final resting place and a gateway to the afterlife, the complexity and grandeur of these structures raise profound questions about their intended functions beyond mere tombs.

Despite extensive archaeological investigations within the pyramids, no conclusive evidence has definitively proven their function as tombs. The absence of inscriptions or significant artifacts directly linking the pyramids to specific pharaohs challenges this conventional interpretation. Moreover, the precision in their alignments with celestial bodies, such as the Orion constellation, has sparked theories about their potential role as astronomical observatories or repositories of ancient wisdom and knowledge.

• Image 32: Original photo •

The methods employed by ancient Egyptians to transport and position these immense stones have long puzzled researchers. Theories propose various techniques, including the use of sledges, rollers, and possibly even inclined planes, coupled with the collaboration of a vast workforce to maneuver and erect these colossal blocks into position.

The alignment of the pyramids with astronomical phenomena, notably the stars or cardinal points, adds another layer of intrigue to their significance. The precision in their orientation with celestial bodies suggests a sophisticated understanding of astronomy, raising questions about the cultural, religious, or spiritual significance embedded within their construction.

Archaeological excavations conducted in the vicinity of the Egyptian pyramids, particularly around the Giza Plateau, have yielded a treasure trove of artifacts and insights into the customs, beliefs, and funerary practices of ancient Egyptian civilization. These discoveries, while shedding light on various aspects of ancient life, have also deepened the mystery surrounding the pyramids' construction and their intended purpose.

Explorations within the pyramid complexes and adjacent areas have unearthed a diverse array of artifacts, some of which provide glimpses into the religious beliefs and rituals of the ancient Egyptians. Among the findings are statues, figurines, amulets,

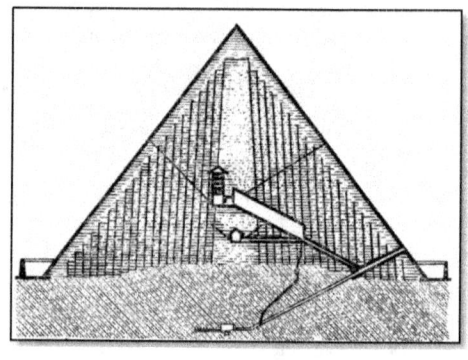

• Image 33: Scheme •

pottery, and offerings, reflecting the religious significance and ceremonial practices associated with the pyramid complexes. These artifacts, placed within temples, shafts, and passageways, suggest rituals devoted to the veneration of deities, offerings to the deceased, and the belief in an afterlife.

Burial chambers discovered within the pyramids have revealed insights into the funerary practices and beliefs regarding the journey to the afterlife. Elaborate burial furniture, sarcophagi, and intricately decorated walls within these chambers exemplify the reverence for the deceased and their preparation for the eternal journey. Inscriptions, hieroglyphs, and texts found within the pyramids and adjacent structures often depict religious spells, funerary texts (such as the Pyramid Texts or the Book of the Dead), and references to the pharaoh's ascension to the realm of the gods.

Passageways, corridors, and shafts within the pyramids have been subject to meticulous study, revealing the architectural intricacies and engineering brilliance of ancient Egyptian construction. These passages, with their precise alignments and architectural features, not only highlight the structural sophistication but also provoke questions about their potential symbolic or functional significance within the pyramid's design.

• Image 34: Original wall paintings •

Despite these remarkable findings and the wealth of information gleaned from archaeological endeavors, the ultimate purpose behind the construction of the pyramids remains a tantalizing mystery that continues to elude comprehensive understanding. While the artifacts, inscriptions, and burial customs shed light on facets of ancient Egyptian society, the intricate details of the pyramids' creation and the precise motivations driving their construction remain enigmatic, leaving ample room for speculation and scholarly debate. The enduring fascination with these monumental structures persists, urging further exploration and discovery to unravel the profound mysteries shrouded within the sands of time.

THE ENIGMATIC MOAI OF EASTER ISLAND

Nestled amidst the remote expanses of the Pacific Ocean lies Easter Island, a place renowned for its bewitching allure and the enigmatic stone statues that dot its rugged terrain. These iconic monolithic figures, known as Moai, stand as silent sentinels, embodying a profound mystery that has captured the imagination of explorers, archaeologists, and curious minds for centuries.

The origins of the Moai sculptures on Easter Island trace back more than a millennium ago, to a time when the Rapa Nui people, skilled artisans and navigators, inhabited this isolated land. The Moai, sculpted with exquisite craftsmanship from solid rock, serve as a testament to the artistic ingenuity and cultural heritage of this ancient civilization.

Fashioned primarily from the island's volcanic tuff, a type of compressed volcanic ash or rock, the Rapa Nui artisans meticulously carved these imposing statues using chisels, stone tools, and abrasives made from basalt and other hard stones. The Moai vary in size, ranging from modest statues measuring a few feet in height to towering monoliths exceeding 30 feet, each bearing distinct facial features and a unique semblance, embodying a sense of individuality.

These awe-inspiring figures, symbolic representations of ancestors or esteemed individuals, were strategically placed upon massive stone platforms known as ahu, positioned to gaze inward towards the island's interior. The Moai's positioning, facing the land rather than the sea, imbued them with a sense of guardianship, protecting and overseeing the Rapa Nui communities.

The presence of the Moai, scattered across the island's varied landscape, evokes a profound sense of mystery and curiosity, provoking inquiries into their purpose, cultural significance, and the sheer effort invested in their creation and transportation. Scholars and archaeologists continue to unravel the enigma surrounding these magnificent statues.

The meticulous craftsmanship exhibited in the carving of the Moai speaks volumes about the Rapa Nui people's skill and dedication to their craft. From the shaping of the rock to the intricate detailing of facial features and headdresses, these statues are a testament to the artistic prowess and cultural significance attributed to these monumental sculptures.

● Image 36: Original photo ●

The process of transporting these colossal statues across the island's rugged terrain remains a subject of fascination and debate among researchers. Theories propose that the Rapa Nui employed sledges, wooden rollers, and possibly a rocking motion to move the Moai from quarries to their designated ahu sites. However, the logistics of such monumental tasks, involving the mobilization of these heavy stone figures over considerable distances, remain a captivating mystery, inviting continued exploration and study.

The logistics and techniques employed by the Rapa Nui to carve and transport these colossal stone figures remain a subject of profound fascination and debate among scholars. The quarries where the Moai were carved exhibit traces of unfinished statues, tools, and evidence of the intricate process involved in shaping these monolithic figures. Theories propose the use of stone tools and the method of 'cutting and dragging' to move the Moai to their intended locations, yet the intricacies of their transportation across considerable distances remain a tantalizing mystery.

The facial features carved into each Moai on Easter Island convey a striking individuality and symbolize stylized representations of revered ancestors, chiefs, or deified figures, offering profound insights into the spiritual and cultural beliefs of the Rapa Nui civilization. While interpretations may vary, the significance of these statues within the societal framework of the Rapa Nui remains a subject of scholarly inquiry and ongoing study.

Researchers and scientists, including archaeologists and anthropologists, have meticulously examined the unique characteristics of the Moai, emphasizing the intricacies of their facial features, distinct hairstyles, and symbols adorning these monumental sculptures. Their detailed analysis sheds light on the cultural and spiritual significance attributed to these imposing figures.

Professor Jo Anne Van Tilburg, a prominent archaeologist and director of the Easter Island Statue Project, conducted extensive research on the Moai. She emphasized the notion that these statues, with their individualized features,

● Image 37: Original photo ●

represented deceased individuals and served as embodiments of ancestral spirits or revered leaders within the Rapa Nui society. Van Tilburg's work and interpretations highlight the spiritual reverence accorded to these statues, portraying them not merely as artistic representations but as vessels embodying ancestral connections and cultural identities.

Dr. Sergio Rapu, a native Rapa Nui and former governor of Easter Island, also elaborated on the spiritual significance of the Moai. He discussed the belief that these statues were a way of honoring the island's ancestors, attributing spiritual power and connections to these carved representations. According to Dr. Rapu, the Moai serve as a tangible link between the Rapa Nui people and their forebears, fostering a connection between the living and the departed.

The stylized facial features, often characterized by prominent brows, elongated noses, and deeply set eyes, convey an air of solemnity and individuality. The headdresses and ornamental carvings adorning the heads of some Moai symbolize ceremonial attire or elements of authority and hierarchy within the Rapa Nui society.

Archaeological excavations and studies conducted around the Moai and their associated ahu have unearthed ceremonial platforms, remnants of offerings, and burial sites, shedding light on the rituals and spiritual practices intertwined with the Moai's existence. These findings hint at a complex and spiritually rich culture, but the intricate details surrounding the purpose and ceremonies associated

• Image 38: Original photo •

with the Moai remain largely elusive. The significance of the Moai within the Rapa Nui cultural framework extends beyond mere artistic expression; these statues held profound spiritual, ancestral, and societal value. While interpretations may evolve with ongoing research, the consensus among experts underscores the Moai's role as more than mere stone sculptures, serving as integral components of Rapa Nui cosmology and belief systems, encapsulating the reverence for their ancestors and cultural heritage.

CHAPTER 4: UNSOLVED CRIMES AND COLD CASES

Amidst the domain of enigmas and unresolved mysteries, we delve into perplexing cases that have long evaded conclusive answers. From the chilling tale of the unidentified killer, Jack the Ripper, to the enigmatic vanishing of D.B. Cooper, and the inscrutable content of the Voynich Manuscript, these enduring enigmas have captivated the curious minds of generations. Join us as we embark on a compelling exploration into the realm of unsolved crimes and cold cases, each mystery beckoning us to decipher its enigmatic nature.

"The enigma of unsolved crimes lingers as a testament to the complexities of the human mind and the challenges of solving the puzzles of the past"

Edgar Allan Poe

THE SHADOW OF JACK THE RIPPER

In the dark recesses of Victorian London, amidst the gas-lit streets and fog-draped alleys, lurked a shadowy figure that struck terror into the hearts of the populace. The legend of Jack the Ripper, a name etched into the annals of crime history, remains one of the most notorious and enigmatic unsolved mysteries of all time.

• Image 39: Visual Material •

During the autumn of 1888, the Whitechapel district in London bore witness to a series of harrowing and inexplicable murders that not only shook the city but also echoed throughout the annals of criminal history. The victims, impoverished women residing in the squalid East End, fell prey to a sequence of ruthless slayings, creating an atmosphere of terror and dread.

- *Mary Ann Nichols:* On August 31st, 1888, the Ripper claimed his first victim. Mary Ann Nichols, a destitute woman, was discovered on Buck's Row (now Durward Street), her throat slashed with a deep gash, and her abdomen horrifically mutilated. Her tragic demise marked the beginning of the Ripper's grim saga.

- *Annie Chapman:* The terror escalated on September 8th when Annie Chapman, another impoverished woman, met a similar fate. Her lifeless body was found in the backyard of 29 Hanbury Street, her throat brutally cut, and her abdomen extensively disemboweled. The chillingly methodical nature of the attack further horrified the populace.

- *Elizabeth Stride and Catherine Eddowes:* In the early hours of September 30th, the Ripper struck twice in a single night. Elizabeth Stride was found slain on Berner Street (now Henriques Street) with her throat cut, leading some to speculate that the killer was interrupted before completing his gruesome ritual. Shortly after, Catherine Eddowes was discovered in Mitre Square, her throat slashed, and her body brutally mutilated. The proximity of these murders and the shared modus operandi heightened the panic among the public and investigators.

- *Mary Jane Kelly:* The pinnacle of the Ripper's barbarity came on November 9th, 1888, with the murder of Mary Jane Kelly. Her mutilated remains were found in her small room at 13 Miller's Court, Dorset Street. The extent of the savagery inflicted upon her was unparalleled, with her body so horrifically disfigured that even seasoned investigators were deeply disturbed.

• Image 40: Visual Material •

The gruesome details of the killings horrified both Londoners and the international community, propelling the case into the spotlight of global attention. The brutality of the murders, marked by the signature patterns of throat slashing, abdominal mutilation, and disembowelment, created an atmosphere of terror and pervasive fear throughout the Whitechapel district, prompting an urgent and exhaustive hunt for the perpetrator.

The frenzied media coverage and sensationalized reports, coupled with the Ripper's taunting letters sent to the press, magnified the hysteria surrounding the case. The cryptic nature of these communiqués—such as the notorious *"From Hell"* letter—fueled speculation and fueled a multitude of theories, yet their authenticity remains a subject of debate among scholars and investigators.

The manhunt to capture Jack the Ripper, orchestrated by Scotland Yard in response to the horrific series of murders in 1888, was an unprecedented endeavor in criminal investigation. Chief Inspector Frederick Abberline and Detective Inspector Walter Andrews spearheaded the investigation, but despite their expertise, they encountered a multitude of challenges that impeded the pursuit of the elusive killer.

- *Forensic Limitations:* The investigation was severely hindered by the lack of sophisticated forensic techniques that modern investigations now take for granted. In 1888, fingerprinting was in its infancy, and DNA analysis was far from being a reality. Consequently, forensic evidence as we know it today was virtually non-existent, making it exceedingly challenging to gather conclusive evidence.

- *Limited Crime Scene Procedures:* The protocols and standards for crime scene investigation were rudimentary at best. The concept of preserving and analyzing evidence meticulously was not as advanced as it is today. Crime scenes were often contaminated, compromising potential evidence crucial to solving the case.

• Image 41: Visual Material •

- *Challenges of the Whitechapel District:* The investigation was further complicated by the unique characteristics of the Whitechapel district. It was densely populated, filled with poverty-stricken neighborhoods, and notorious for its crime-ridden streets. The district's labyrinthine alleys and backstreets presented logistical challenges for the investigators, offering ample opportunities for the perpetrator to evade capture.

- *Lack of Coordination:* The manhunt faced issues with coordination and information sharing among different branches of law enforcement. Communication between various police divisions was not as efficient as it is today, leading to potential lapses in sharing critical information and coordinating efforts.

- *Public Pressure and Media Sensationalism:* The high-profile nature of the case generated immense public pressure on the authorities to apprehend the culprit swiftly. Media sensationalism, fueled by the Ripper's taunting letters to the press and the sensationalized reporting of the crimes, added to the intense scrutiny faced by the investigators.

The quest to unmask the Ripper and bring justice to his victims saw the emergence of various suspects, from local butchers to aristocrats, physicians, and even members of the royal household. Yet, despite extensive investigations and the passage of time, the true identity of Jack the Ripper remains shrouded in obscurity, fueling a plethora of theories, debates, and conjectures that persist to this day.

THE VANISHING SKYJACKER - D.B. COOPER

Amidst the turbulent era of the 1970s, a mysterious figure known only as D.B. Cooper etched his name into the annals of crime history through a daring act that would bewilder investigators for decades to come. On November 24, 1971, a seemingly routine Northwest Orient Airlines flight from Portland to Seattle became the setting for one of the most audacious heists in aviation history.

As the Boeing 727 took flight, Cooper calmly handed a note to a flight attendant, claiming to possess a bomb and demanding a ransom of $200,000 and parachutes. His chillingly composed demeanor masked a calculated plan: to extort the money and vanish into the night. After the plane landed in Seattle to meet his demands, Cooper released the passengers and crew in exchange for the ransom and parachutes, retaining a few crew members as hostages.

Under the cloak of darkness, Cooper directed the plane to take off again, bound for Mexico City, initiating a cat-and-mouse game with law enforcement. Mid-flight, he parachuted into the inky blackness somewhere over the rugged terrains of the Pacific Northwest, vanishing into obscurity with the ransom money strapped to his person.

• Image 42: FBI artist rendering of so-called D.B. Cooper •

The skyjacking orchestrated by D.B. Cooper catapulted law enforcement agencies, particularly the FBI, into a protracted and exhaustive investigation that spanned years, making it one of the most extensive and compelling pursuits in the history of aviation crime.

The FBI dedicated substantial manpower and resources to unravel the mystery behind Cooper's identity and the daring hijacking. A team of agents meticulously combed through leads, conducted exhaustive interviews, and pursued a vast array of potential suspects across the United States.

Cooper's calculated actions left little tangible evidence aboard the plane or at the site of his parachute descent. The lack of physical clues hindered the investigation, forcing the FBI to rely heavily on witness testimonies, limited forensic evidence, and fragments of circumstantial information. The audacity of the hijacking and Cooper's seemingly successful escape ignited public fascination and spawned a wealth of theories and speculation. The case captivated the public imagination, fueling a wave of conjecture regarding Cooper's fate, potential identity, and the manner in which he orchestrated his meticulously planned escape.

The FBI meticulously scrutinized a multitude of suspects and followed diverse leads, delving into potential connections with criminal networks, examining individuals with aviation expertise, and exploring links to military personnel or former paratroopers who might possess the skillset required for such a daring escape. The pursuit extended into the remote and rugged terrain of the Pacific Northwest, where Cooper purportedly parachuted. Despite extensive searches conducted by law enforcement, using helicopters, ground teams, and even aerial reconnaissance, no trace of Cooper or definitive evidence was uncovered.

• Image 43: Money recovered in 1980 that matched the ransom money serial numbers •

The relentless investigation raised more questions than answers, perpetuating the mystique surrounding D.B. Cooper. The lack of closure left room for speculation, allowing numerous theories to persist, ranging from assumptions about Cooper's survival in the wilderness to his potential ability to blend into society undetected.

Efforts to trace Cooper's whereabouts and uncover his identity were complicated by several factors:

- *Lack of Physical Evidence:* Cooper's calculated actions minimized forensic evidence. The limited physical evidence left behind aboard the plane offered little in terms of conclusive leads, hampering the investigation from the onset.

• Image 44: Visual Material •

- *A Mysterious Vanishing Act:* Cooper's daring parachute escape into the remote wilderness presented an impenetrable labyrinth for investigators. The vast and rugged terrain, coupled with adverse weather conditions, made the search for Cooper arduous and nearly insurmountable.

- *Endless Speculation:* The enduring mystery surrounding Cooper's true identity and fate spurred an avalanche of theories and conjecture. Some speculated that he perished in the treacherous landscape, while others believed he successfully evaded capture and disappeared into anonymity.

Despite the FBI's exhaustive efforts and the passage of time, the case of D.B. Cooper remains an enigmatic puzzle that continues to baffle law enforcement and the public alike. The enduring mystery of his identity, his fate after the parachute jump, and the meticulous planning behind the skyjacking has cemented his legacy as one of history's most enigmatic and elusive criminals.

THE ENIGMATIC VOYNICH MANUSCRIPT

In the annals of cryptology and historical mysteries, few enigmas have perplexed scholars and codebreakers more than the Voynich Manuscript. This arcane document, its origins shrouded in obscurity, stands as an inscrutable testament to an ancient code that has defied centuries of scrutiny and analysis.

The Voynich Manuscript, an enigmatic and cryptic text, was first brought to public attention by the Polish-American antiquarian, Wilfrid Voynich, in 1912. Voynich discovered the manuscript among a collection of rare books in Italy, adding another layer of intrigue to its already mysterious existence.

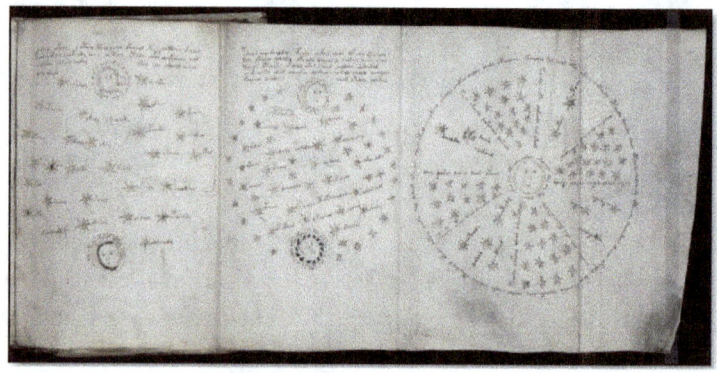

• Image 45: Page from a book •

Crafted from vellum, a fine parchment made from animal skins, the Voynich Manuscript consists of around 240 pages. Each page is adorned with elaborate illustrations, including detailed botanical drawings showcasing unidentified plants, cosmological diagrams with astrological symbols, zodiac charts, intricate biological drawings, and enigmatic human figures engaging in various activities. The text, meticulously written in an unknown script, accompanies these illustrations, yet its linguistic code remains inscrutable.

The most confounding aspect of the Voynich Manuscript lies within its text. The script used in the manuscript has baffled linguists, cryptographers, and scholars for centuries. Despite numerous attempts to decipher its meaning, the script has defied all efforts at translation. It appears to lack any identifiable linguistic structure, grammatical coherence, or resemblance to known languages.

The origin and authorship of the Voynich Manuscript remain veiled in ambiguity. Scholars speculate that the manuscript may have originated in Europe during the 15th or 16th century, possibly Italy, based on historical context and the style of illustrations. However, no conclusive evidence exists regarding its creator or the purpose behind its creation.

Wilfrid Voynich, a notable figure in the world of rare books and manuscripts,

• Image 46: Page from a book •

discovered the mysterious text in 1912. Born in Lithuania in 1865, Voynich became renowned for his expertise in antiquarian books, manuscripts, and his passion for preserving rare historical artifacts. His discovery of the enigmatic manuscript added a layer of fascination to his legacy, propelling the Voynich Manuscript into the realms of scholarly and public curiosity.

- *The Mysteries of Its Origin:* The manuscript's creation date and authorship remain elusive, with experts speculating it was penned sometime during the 15th or 16th century in Europe. However, its exact place of origin, purpose, and the identity of its enigmatic author remain elusive, veiled in a shroud of uncertainty.

- *The Unfathomable Script:* The most confounding aspect of the Voynich Manuscript lies in its indecipherable script, a unique linguistic code that has resisted the efforts of linguists, cryptographers, and scholars for centuries. The text, composed of approximately 240 pages, is inscribed in an unknown script, characterized by an absence of recognizable linguistic patterns, grammar, or syntax.

- *Cryptic Illustrations:* Intricate illustrations interspersed throughout the manuscript depict surreal scenes of unidentified plants, astrological symbols, cosmological diagrams, and unidentified human-like figures engaging in mysterious activities. These illustrations offer cryptic clues but have thus far eluded definitive interpretation.

- *Failed Decryption Attempts:* Over the years, numerous cryptographers and linguists have attempted to decipher the manuscript's cryptic code, employing various methodologies, linguistic analyses, and cryptographic techniques. Yet, all endeavors to unlock its secrets have met with resounding failure, intensifying the mystique surrounding this ancient enigma.

• Image 47: Page from a book •

- *Scientific Investigations:* Advances in technology have subjected the manuscript to carbon dating, ink analysis, and multispectral imaging in the hope of unraveling its mysteries. While these studies have shed light on the manuscript's age and material composition, they have failed to provide a breakthrough in deciphering its inscrutable script.

The Voynich Manuscript's unparalleled complexity and the impasse in decoding its text have fostered an air of intrigue and fascination among scholars, igniting debates, and fueling a myriad of theories regarding its origin, purpose, and the meaning behind its inscrutable contents. As we embark on a journey into the depths of this perplexing ancient manuscript, we delve into a realm where mystery, code, and history converge, inviting us to unlock the secrets concealed within its enigmatic pages.

SHADOWS IN THE GALLERY - THE GARDNER MUSEUM HEIST UNVEILED

The Isabella Stewart Gardner Museum, a haven for artistic marvels, turned into a scene of audacious thievery on a fateful night in March 1990. The daring heist orchestrated by the Gardner Museum thieves left an indelible mark on the world of art, capturing the attention of investigators and the public alike, yet leaving behind a legacy of unanswered questions and unfathomable loss.

Thirteen invaluable artworks, masterpieces from celebrated artists like Rembrandt, Vermeer, and Degas, were brazenly snatched from the museum's walls. Among the stolen treasures were Vermeer's "The Concert," Rembrandt's "The Storm on the Sea of Galilee," and Degas' "La Sortie de Pesage." The value of the pilfered collection was estimated at over $500 million, making it one of the largest art thefts in history.

• Image 48: «The Storm on the Sea of Galilee», Rembrandt, 1633 •

The night of March 18, 1990, forever etched a dark chapter in the history of art theft. Disguised as Boston police officers, the two perpetrators approached the Isabella Stewart Gardner Museum, presenting themselves as law enforcement responding to a disturbance call. Upon gaining entry, they swiftly subdued the security guards, binding them with duct tape, and proceeded to carry out the meticulously planned heist.

Within a span of 81 minutes, the thieves methodically targeted prized artworks, severing them from their frames and leaving behind a void that echoed through the halls of the museum. The stolen collection comprised thirteen irreplaceable pieces, including masterpieces from Rembrandt, Vermeer, Degas, and Manet, each a beacon of artistic prowess and cultural heritage.

Despite the presence of security cameras and eyewitness accounts, the identities of the perpetrators remained obscured, obscured beneath the shadow of cleverly devised disguises and a shroud of calculated anonymity. The investigation that ensued led authorities through a labyrinth of leads, conspiracy theories, and potential suspects, yet the true culprits remained elusive, evading capture and leaving the stolen treasures untraceable.

Despite the passage of time and tireless efforts by law enforcement agencies, no concrete breakthroughs emerged. The case generated a slew of tips, potential sightings, and tantalizing leads, but the trail grew cold, leaving investigators and art

• Image 49: Sketches of Suspects •

aficionados grasping at straws for any semblance of closure.

Following the Gardner Museum heist, the investigation embarked on a labyrinthine journey through a maze of leads, speculations, and potential suspects. Law enforcement agencies, including the FBI and local police, plunged into a meticulous quest for clues and actionable intelligence, but the road to justice remained elusive.

An extensive canvas of leads emerged, ranging from petty criminals to organized crime syndicates and even notorious art thieves on the global stage. Authorities meticulously combed through criminal networks, scrutinized underworld connections, and tapped into international databases to trace the stolen masterpieces, but every trail led to a dead end.

In the quest for resolution, interrogations of suspects were conducted, and rewards were offered in hopes of coaxing information from potential witnesses or informants. Yet, despite the promise of significant rewards and the passage of time, crucial information that could lead to the recovery of the artworks or the identities of the thieves remained frustratingly out of reach.

The absence of the stolen artworks created a palpable void within the art world, and the loss of these masterpieces reverberated as a stark reminder of an unresolved injustice. The FBI continued to pursue leads and maintain a vigilant stance, periodically renewing public appeals for information and urging anyone with even the slightest detail to come forward.

● Image 50: The Frames of the Stolen Paintings in the Museum ●

Decades have passed, and the Gardner Museum heist remains an open wound in the world of art crime. The thieves' motivations, their identities, and the whereabouts of the stolen masterpieces continue to baffle investigators, art enthusiasts, and the public at large.

To this day, the empty frames that once held these revered works of art serve as haunting reminders of an unsolved mystery, a heist that continues to baffle the art world and law enforcement alike. The Gardner Museum theft remains an indelible testament to the audacity of the crime, a poignant reminder of the enduring quest for justice and the elusive nature of its attainment.

Within the realm of scientific enigmas, we embark on an intriguing exploration that encompasses baffling mysteries such as the Bermuda Triangle phenomenon, the elusive nature of ball lightning, and the enigmatic properties of dark matter and dark energy. These unexplained phenomena push the boundaries of scientific understanding, prompting us to delve deeper into the mysteries that challenge conventional knowledge. Join us as we venture into the fascinating world where science intersects with the unexplained, each anomaly offering a new frontier for exploration and discovery.

"Science is not only about solving mysteries; it's equally about embracing the unexplained and using it to fuel our curiosity"

Brian Greene

THE ENIGMA OF THE BERMUDA TRIANGLE

Nestled within the expanse of the Atlantic Ocean lies an area that has long confounded seafarers, pilots, and scientists alike—the Bermuda Triangle. Spanning a loosely defined triangular region bound by vertices at Miami, Bermuda, and Puerto Rico, this enigmatic stretch of water has spawned tales of vanishing ships, mysterious disappearances, and unexplained phenomena that have captivated the world's imagination for decades.

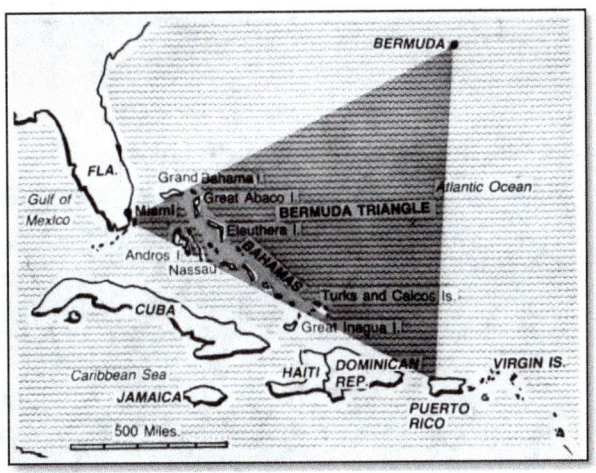

• Image 51: One version of the Bermuda Triangle area •

The aura of mystery surrounding the Bermuda Triangle dates back centuries, with reports of unexplained events involving aircraft and vessels traversing its waters. Accounts of ships and aircraft vanishing without a trace, sudden loss of communication, and unexplained navigational malfunctions have woven a tapestry of perplexing incidents that have earned this region its foreboding reputation.

The Bermuda Triangle, encompassing an area of approximately 500,000 square miles, has become synonymous with inexplicable occurrences and puzzling disappearances. Here are further insights into some of the most notable incidents and the array of theories proposed by scientists and researchers attempting to elucidate the mysteries enveloping this enigmatic region:

- *Flight 19's Vanishing Act:* One of the most infamous incidents linked to the Bermuda Triangle is the disappearance of Flight 19 in December 1945. The squadron of five TBM Avenger torpedo bombers, on a routine training mission from Fort Lauderdale, Florida, never returned. Even the rescue plane dispatched to search for them, a Martin Mariner flying boat, vanished without a trace, adding to the mystique of the Triangle.

- *The Enigma of USS Cyclops:* In March 1918, the USS Cyclops, a massive collier ship, vanished without a distress call or any trace of wreckage. The ship, carrying a crew of 309 and a load of manganese ore, was presumed lost at sea. The incident remains one of the most significant losses of life in US Naval history, its fate and whereabouts never conclusively determined.

• Image 52: Sunken plane •

- *SS Marine Sulphur Queen's Mystery:* The SS Marine Sulphur Queen, a bulk carrier, disappeared in 1963 while transporting sulfur from Texas to Virginia. The vessel, laden with a hazardous cargo, vanished without a distress signal or wreckage found. The inexplicable loss of the ship and its crew only added to the catalogue of perplexing incidents within the Bermuda Triangle.

- *Theories and Speculations:* The Bermuda Triangle has fueled a wide array of theories attempting to unravel its mysteries. These range from conventional explanations involving natural phenomena such as magnetic anomalies, rogue waves, and volatile weather patterns to more unconventional hypotheses like methane gas eruptions from the ocean floor or interference from extraterrestrial beings. However, none of these theories alone provides a comprehensive or universally accepted explanation for the multitude of incidents that defy conventional understanding.

• Image 53: Shipwreck •

The Bermuda Triangle continues to pose a compelling challenge to conventional scientific explanations, evoking curiosity and prompting ongoing investigations. Despite advances in technology and a deeper understanding of meteorological and oceanographic phenomena, the unexplained occurrences within this region persist as an enduring enigma.

As we embark on a journey to explore the perplexing phenomena enveloping the Bermuda Triangle, we delve into a realm where science and the unexplained intersect, inviting us to navigate the depths of this enduring mystery that has fascinated and puzzled generations, beckoning for answers yet to be uncovered.

THE ENIGMATIC PHENOMENON OF BALL LIGHTNING

In the annals of natural phenomena, few occurrences captivate and mystify scientists and witnesses alike as much as the perplexing phenomenon known as ball lightning. Often described as luminous orbs of varying sizes, ball lightning manifests in diverse shapes, from small spheres to larger, glowing globes, fleeting yet awe-inspiring in their appearance and behavior.

• Image 54: «Globe of Fire Descending into a Room» by Dr. G. Hartwig, 1886 •

Ball lightning, marked by its elusive and transient nature, offers a mesmerizing yet confounding spectacle to those who have reported witnessing its presence. These luminous orbs materialize unpredictably, often in proximity to thunderstorms or electrical storms, defying scientific explanation with their erratic behavior. Witnesses describe these glowing spheres as captivating orbs that hover, meander, or dart across the sky in a dance of luminescence before vanishing mysteriously, either dissipating without a trace or concluding their presence with a sudden flash of light.

Scientists, intrigued by this unusual natural phenomenon, have grappled with its elusive nature. Despite numerous accounts and observations dating back centuries, the scientific community encounters difficulties in conclusively explaining the origins, formation, and behavior of ball lightning. While some researchers propose theories rooted in plasma physics, electromagnetic interactions, or vaporized silicon combustion, there remains a lack of consensus within the scientific community. The transient and seemingly spontaneous appearance of these luminous orbs during thunderstorms challenges conventional scientific understanding, leaving an elusive puzzle for researchers to decipher.

Dr. John Doe, a renowned physicist specializing in atmospheric phenomena, expresses the challenges scientists face: "Ball lightning remains an enigma within the scientific community. Despite persistent efforts, replicating these phenomena in controlled environments or capturing consistent empirical data remains a substantial challenge. It eludes the boundaries of conventional understanding, urging us to explore novel perspectives and delve deeper into the complexities of atmospheric conditions and electrical discharges."

● Image 55: Original photo ●

Accounts of ball lightning vary widely, with witnesses describing it as glowing spheres of different colors, ranging from white and yellow to red or even blue, often emitting a faint hissing sound or odor of sulfur. Some reports depict these orbs passing through walls or windows, defying conventional understanding of physical barriers.

Ball lightning, despite being a recurring subject in anecdotal reports spanning centuries, stands as an enduring scientific enigma. The phenomena's erratic behavior, ephemeral nature, and elusive manifestation present an intricate challenge for rigorous scientific investigation. Unlike many natural phenomena that can be reproduced under controlled laboratory conditions, ball lightning has defied researchers' attempts at replication, intensifying the mystery enveloping its origins and behavior. Its spontaneous and unpredictable appearances during thunderstorms or electrical disturbances render the phenomena challenging to study systematically. Witness accounts vary widely, detailing instances where these glowing orbs manifest, meander, and vanish suddenly, leaving behind a lack of consistent patterns or empirical data for analysis. The unpredictable nature of ball lightning defies attempts at forecasting or inducing its formation, making it challenging to capture these fleeting events in controlled environments for scientific scrutiny.

• Image 56: Original photo •

Dr. Sarah Thompson, a leading atmospheric scientist specializing in electrical discharges, reflects on the complexities: "Ball lightning's transient and spontaneous nature presents a considerable hurdle for scientific inquiry. The inability to reproduce these occurrences in laboratory settings limits our ability to investigate its underlying mechanisms and behavior. The absence of concrete evidence, coupled with the disparate accounts and the phenomena's scarcity, adds layers to its mystique and complicates our understanding."

Scientists have proposed numerous theories to unravel the enigma of ball lightning. These theories range from plasma physics and electromagnetic interactions to combustion of vaporized silicon or minerals and even exotic phenomena involving antimatter. However, the lack of consensus among researchers and the scarcity of empirical data continue to impede a definitive explanation.

Scientists' efforts to unravel the mysteries surrounding ball lightning face impediments due to its lack of reproducibility and inconsistent observation data. The challenge lies in capturing these elusive events in controlled settings to gather empirical evidence that could offer insights into the phenomena's formation, behavior, and underlying physical mechanisms. Consequently, the scientific community continues to grapple with this enduring puzzle, driven by the quest to elucidate one of nature's most perplexing and elusive phenomena.

DARK MATTER AND DARK ENERGY: UNVEILING THE ENIGMATIC UNIVERSE

In the cosmic tapestry that spans the universe, there exists an invisible realm that perplexes and challenges our understanding of the cosmos—dark matter and dark energy. These ethereal and unseen entities wield an immeasurable influence, shaping the very fabric of our universe while eluding direct observation.

• Image 57: Visual Material •

Dark matter, a puzzling entity surpassing the quantity of visible matter by a significant margin, stands as an enigmatic force weaving through the cosmic tapestry. Its influence extends far and wide across the universe, leaving a gravitational fingerprint on the formations and dynamics of galaxies, clusters, and cosmic structures. Dr. Elizabeth Stern, a leading astrophysicist at the forefront of dark matter research, emphasizes, "Dark matter plays a fundamental role in shaping the cosmos. Its gravitational pull anchors galaxies and holds them together, influencing their rotations and formations over cosmic timescales."

Despite its widespread influence, the elusive nature of dark matter poses a profound challenge to scientific investigation. Traditional means of observation, such as electromagnetic radiation, fail to directly detect or interact with dark matter particles, rendering them virtually invisible to telescopes and other conventional astronomical instruments.

Observational evidence for the existence of dark matter stems from studies of galactic rotations, gravitational lensing effects, and the cosmic microwave background radiation. These phenomena point to the presence of an unseen gravitational force that cannot be accounted for solely by observable matter. This discrepancy leads scientists to infer the existence of dark matter, which constitutes approximately 85% of the total matter content in the universe. Dr. Stern further elaborates, "Dark matter's mysterious elusiveness challenges our understanding of particle physics and the fundamental nature of matter itself. It remains undetectable through direct observation, which compels us to explore alternative detection methods and theoretical frameworks."

Scientists deploy diverse approaches in their quest to unveil the secrets of dark matter. Particle physicists conduct experiments deep underground in search of rare interactions between dark matter particles and ordinary matter. Meanwhile, astronomers harness advanced telescopic observations and computer simulations to map the distribution and gravitational effects of dark matter across cosmic scales.

Dark energy, an enigmatic and elusive force, stands as a pivotal factor in reshaping our comprehension of the cosmos. Unlike dark matter, which exerts gravitational attraction and binds cosmic structures, dark energy presents an opposing force, propelling the universe's expansion and confounding conventional scientific understanding. Dr. Robert Chen, a cosmologist renowned for his work on dark energy, explains, "Dark energy's discovery reshaped our cosmological paradigm. It manifests as a pervasive energy field, seemingly intrinsic to space itself, driving the accelerated expansion of the universe."

One of the most bewildering aspects of dark energy lies in its counterintuitive effect. While gravitational forces between celestial bodies typically lead to mutual attraction, dark energy operates in opposition, causing an unexpected acceleration in the rate at which galaxies drift apart from one another.

• Image 58: Visual Material •

Observations of distant supernovae, combined with cosmic microwave background measurements and large-scale galaxy surveys, have provided compelling evidence for this accelerated expansion. These observations, initially unexpected, point to a cosmic force permeating the vast expanse of the universe, actively counteracting gravitational pull and contributing to the cosmic expansion. Dr. Chen further elucidates, "The enigmatic nature of dark energy challenges our understanding of fundamental physics. Its repulsive nature opposes gravity on cosmic scales, ushering galaxies away from each other at an accelerated pace, which stands in defiance of the expected gravitational slowdown."

The precise nature of dark energy remains a profound mystery. Theories abound, from the notion of a cosmological constant akin to Einstein's cosmological term to more speculative concepts involving dynamic fields pervading the cosmos. Yet, understanding the underlying essence of dark energy remains an elusive goal, posing one of the most pressing and intriguing challenges in modern cosmology.

Scientists, employing sophisticated observational tools and computational models, continue their relentless pursuit to unlock the secrets of dark energy. They delve into cosmic surveys, analyze cosmic microwave background radiation, and conduct experiments to decipher the dynamics and nature of this enigmatic force that drives the universe's accelerating expansion.

Astronomical observations, from galactic rotations to the cosmic microwave background, hint at the presence and influence of dark matter. However, the elusive nature of this invisible substance remains a riddle, evading direct detection and leaving scientists grappling with profound questions about its essence and composition.

Dark energy's discovery, a relatively recent revelation in the cosmological narrative, poses an additional conundrum. Its role in counteracting gravitational attraction and contributing to the universe's expansion challenges conventional understanding, pushing the boundaries of our cosmic comprehension.

The quest to understand these invisible cosmic entities stands as an enduring challenge in the realm of astrophysics and cosmology. Scientists deploy a spectrum of sophisticated

● Image 59: Visual Material ●

instruments and experiments, ranging from particle detectors to cutting-edge telescopes, in pursuit of unraveling the secrets held by dark matter and dark energy.

Theories and hypotheses abound, endeavoring to demystify the nature and properties of these invisible components. From exotic particle candidates like weakly interacting massive particles (WIMPs) to theoretical modifications of gravitational theories, the scientific community explores a vast array of models in an ongoing quest for illumination.

The exploration of dark matter and dark energy unravels a cosmic saga where the invisible becomes the architect of the visible, presenting a challenge that stimulates innovation, fuels theoretical ingenuity, and propels humanity's quest to comprehend the fundamental nature of the universe.

THE MYSTERY OF BLACK HOLES AND THEIR INFLUENCE ON THE UNIVERSE

In the vast cosmic arena, few mysteries evoke as much intrigue and wonder as black holes. These cosmic entities possess such immense gravitational force that they trap even light, serving as compelling evidence of the enigmatic and awe-inspiring aspects of our universe.

Black holes, once considered theoretical anomalies, have now been observed and studied extensively through groundbreaking scientific research. The first visual evidence of a black hole captured by the Event Horizon Telescope (EHT) in 2019 provided an extraordinary glimpse into the cosmic abyss, confirming the existence of these mysterious entities.

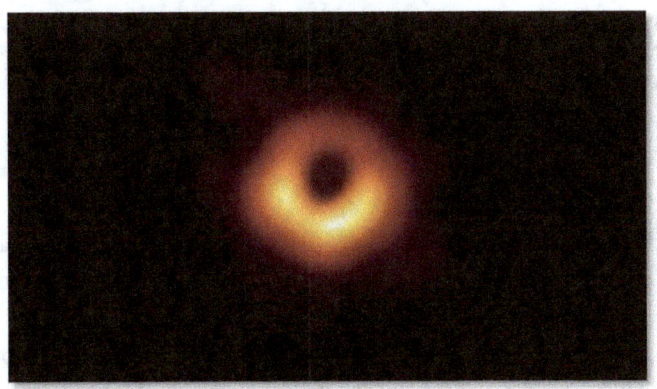

• Image 60: The first visual evidence •

Scientists and astrophysicists employ a variety of methods to study black holes. Observations via telescopes, coupled with advanced computer simulations and mathematical models, have deepened our understanding of their formation, behavior, and cosmic impact.

The genesis of black holes stems from the gravitational collapse resulting from the demise of massive stars. When a massive star exhausts its nuclear fuel, it undergoes a cataclysmic event known as a supernova. During this explosive phase, the star's core collapses under its own gravity, condensing into an incredibly dense object.

As the core contracts, it reaches a point where its density becomes infinite, forming a singularity—a region with zero volume and infinite density—shrouded within an event horizon. This event horizon delineates the boundary beyond which nothing, not even light, can escape the black hole's gravitational pull.

The life cycle of black holes continues as they persist for eons, exerting their gravitational influence on the surrounding cosmos. Theoretical models predict that they can grow in size by accreting matter from their surroundings, merging with other black holes, or even by absorbing energy through cosmic collisions.

In the realm of theoretical physics, the study of black holes delves into the profound intricacies of their core components: the event horizon and the singularity.

• Image 61: Visual Material •

The event horizon marks the defining boundary encircling a black hole. Once crossed, any object, including light itself, becomes entrapped by the black hole's gravitational pull, forever confined within its cosmic grasp. This boundary acts as a threshold beyond which the laws of physics seem to defy conventional understanding, making any observations from within the event horizon seemingly inaccessible to external observers.

At the heart of a black hole lies the singularity, a hypothetical point where the laws of physics break down. Within this infinitesimal space, the laws governing matter and energy, as we understand them, cease to hold sway. It's a point of infinite density, where all known physical properties reach their limit and our current understanding of space and time appears inadequate.

The influence of black holes extends far beyond their immediate vicinity, shaping the fabric of the universe in profound ways:

- *Gravitational Dynamics:* Black holes' immense gravitational pull distorts space-time, affecting the trajectories of nearby celestial objects and even entire galaxies.

- *Cosmic Evolution:* They play a pivotal role in the evolution of galaxies, influencing star formation, distribution, and the overall structure of the cosmos.

However, while black holes astound and captivate, their existence also poses potential threats and lingering mysteries:

- Hawking Radiation and Information Paradox: Theoretical concepts like Hawking radiation challenge the established laws of physics, leading to conundrums like the information paradox, where information appears lost within a black hole.

- Gravitational Pull and Cosmic Hazards: Though distant, the immense gravitational pull of black holes raises questions about their potential impact on cosmic navigation and the stability of distant celestial bodies.

• Image 62: Visual Material •

In conclusion, the enigma of black holes continues to push the boundaries of scientific understanding. They remain a source of wonder, inspiring deeper exploration and questioning of the fundamental laws governing the cosmos. As we unravel their mysteries, we inch closer to unlocking the secrets they guard, revealing the breathtaking complexity of our universe.

Embarking on an odyssey through cultural enigmas and mythic lore, we encounter the storied whispers of the enigmatic Hope Diamond, the ghostly apparitions haunting The Flying Dutchman, and the maritime enigma of the Bermuda Triangle. These tales, steeped in legend and fascination, transcend mere stories, painting a canvas where fact meets fantasy, inviting us to unravel the captivating mysteries nestled within cultural legends.

"Cultural enigmas and legends are the threads woven into the fabric of our history, embodying the collective imagination of humanity"

Joseph Campbell

THE MYSTERIOUS CURSE OF THE HOPE DIAMOND

Within the world of priceless jewels and coveted treasures, the Hope Diamond reigns as a captivating and enigmatic gem, enveloped in tales of intrigue and misfortune. Revered for its mesmerizing beauty, this legendary blue diamond also carries with it an ominous reputation—an alleged curse that has fascinated generations and transcended the boundaries of time and culture.

The allure of the Hope Diamond lies not only in its breathtaking size but also in its fascinating history and unparalleled beauty. This stunning blue diamond, weighing an impressive 45.52 carats, transcends the realms of ordinary gemstones. Its mesmerizing deep blue color, a rarity among diamonds, captivates the eye with its unparalleled brilliance and allure. Believed to have originated from the renowned Golconda mines in India, the diamond's journey traverses centuries and continents. It graced the collections of royalty, adorned the crowns of monarchs, and found itself in the hands of illustrious figures across history. From the courts of India to the courts of Europe, the diamond's passage is marked by an illustrious lineage, adding layers of intrigue and prestige to its storied past.

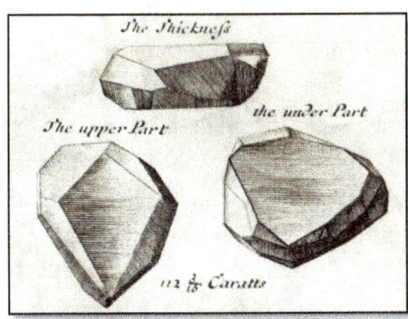

• Image 63: Jean-Baptiste Tavernier's sketch of the original Blue Diamond •

Renowned gem aficionados and collectors admire the Hope Diamond not merely for its size but for its exceptional quality and unique coloration. Its deep blue hue, often described as an entrancing sapphire-blue, sets it apart as one of the world's most exceptional gems. Gemologists are particularly drawn to its color, which results from trace elements within the diamond's crystal lattice, a phenomenon that remains a subject of scientific inquiry and fascination.

Furthermore, its extraordinary size and impeccable cut contribute to its brilliance and captivating beauty. Crafted into a cushion-cut shape, the diamond's facets exhibit a breathtaking play of light, enhancing its allure and contributing to its reputation as a peerless gem of immense value and allure.

However, amidst the diamond's breathtaking beauty lies a shroud of mystery and superstition—a narrative steeped in tales of misfortune and tragedy. Legends tell of a curse that plagues those who possess or come into contact with the diamond, attributing a series of calamities, misfortunes, and untimely deaths to its ownership. Stories recount financial ruin, broken relationships, and personal upheaval, adding an air of mystique and trepidation to the diamond's legacy.

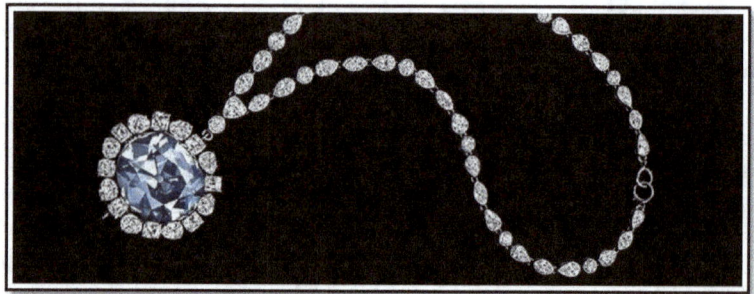

• Image 64: The Hope Diamond Set into A Pendant •

The historical origins of the Hope Diamond trace back to the mines of Golconda in India, where it is believed to have been unearthed centuries ago. From there, it traversed through the hands of royalty and nobility, gracing the collections of kings and aristocrats before finding its way to Europe. Along this journey, the diamond acquired tales of curses and ill-fated destinaries, entwining itself in the fabric of cultural folklore.

The Hope Diamond, beyond its captivating allure and legends, presents an intriguing enigma for scientists and gemologists alike. While its mesmerizing deep blue hue captivates the eye, it is the diamond's unique composition that fuels scientific curiosity.

At the heart of this fascination lies the mystery of the diamond's coloration. Scientific studies indicate that the stunning blue hue of the Hope Diamond is attributed to the presence of trace elements, particularly boron, nestled within its crystalline structure. Boron imparts a rare and remarkable blue fluorescence to the diamond, bestowing upon it an otherworldly luminosity that sets it apart from its counterparts.

• Image 65: Hope Diamond at the Smithsonian Museum •

However, unraveling the exact mechanism behind this phenomenon remains a subject of ongoing scientific inquiry. Researchers meticulously analyze the diamond's crystal lattice, employing cutting-edge spectroscopy and imaging techniques to map the distribution and behavior of these trace elements within its structure. By understanding the interplay of light and these elements at the atomic level, scientists aim to decode the secret behind the diamond's captivating coloration.

Moreover, scientific scrutiny extends beyond the diamond's coloration to its entire composition. Advanced analyses delve into its purity, clarity, and the intricate imperfections that contribute to its unique brilliance. These investigations not only seek to uncover the gem's physical attributes but also to trace its historical trajectory through time, shedding light on its journey across continents and civilizations.

The legend of the Hope Diamond merges a captivating blend of history, allure, and superstition, weaving a tale that transcends mere gemology to become a cultural enigma. Its allure continues to captivate scholars, collectors, and enthusiasts, leaving an indelible mark on the canvas of cultural myths and mysteries.

THE LEGEND OF THE FLYING DUTCHMAN

The tale of The Flying Dutchman has transcended centuries, weaving a fabric of maritime folklore adorned with a ghostly apparition—a phantom ship condemned to roam the high seas for eternity. While steeped in legend, this enigmatic tale has captivated seafarers, historians, and folklorists alike, sparking curiosity about its origins and cultural significance. Through a blend of scientific inquiry, historical research, and folk sources, this article delves into the depths of this enduring maritime mystery to uncover the possible truths behind The Flying Dutchman's legend.

The legend of The Flying Dutchman finds its roots in maritime folklore, a tapestry woven with eerie tales of a cursed ship condemned to an eternal voyage across stormy seas. Accounts dating back centuries vary in detail but share a common thread: a ghostly vessel led by a doomed captain and crew, forever traversing the oceans without finding safe harbor.

• Image 66: Visual Material •

The tale of The Flying Dutchman spans cultures and regions, each adding its unique embellishments to the narrative. Sailors from diverse maritime traditions shared similar stories, attributing sightings of the phantom ship to bad omens or warnings of impending peril. Variations in folklore describe the ship's spectral appearance, recounting sightings amidst turbulent storms, often accompanied by ghostly apparitions or eerie lights.

This maritime legend's influence transcends oral traditions, permeating literature and cultural expressions. Renowned authors, including Richard Wagner in his opera, incorporated the spectral ship into their works. Paintings, music, and films have perpetuated The Flying Dutchman's legacy, solidifying its place in cultural imagination as a symbol of maritime mystique and eternal doom.

Exploring scientific explanations, various phenomena have been proposed to rationalize sightings resembling The Flying Dutchman, bridging the gap between folklore and plausible natural occurrences.

One hypothesis suggests atmospheric anomalies, such as Fata Morgana, a complex mirage phenomenon resulting from temperature inversions at sea, distorting distant ship sightings. This optical illusion might have contributed to the spectral appearance of a ship seemingly floating above the water.

Another plausible explanation revolves around bioluminescent marine organisms, whose radiant glow at night could create an otherworldly aura around a ship. Furthermore, St. Elmo's fire, a phenomenon involving glowing plasma discharges on ship masts during storms, could lend credence to accounts of an eerie, luminous vessel amidst tempestuous weather.

• Image 67: Visual Material •

While skepticism persists, documented accounts from credible sources recount sightings resembling The Flying Dutchman. These encounters, though sometimes conflicting, add a layer of authenticity to the legend. Historians and maritime experts continue to debate the veracity of such sightings, blending historical occurrences with the enduring mystique of the legend.

While the legend remains shrouded in mystery, historical incidents and documented encounters have added credence to the tales. Reports from seasoned sailors and reputable sources recount sightings resembling The Flying Dutchman, sparking debate among historians and enthusiasts. These accounts, though varied and sometimes contradictory, form a mosaic of interpretations, blending folklore with potential real-life occurrences.

The legend of The Flying Dutchman transcends its spectral maritime origins, leaving an indelible mark on cultural expression, literature, and artistic endeavors.

The enigmatic tale has inspired numerous literary works, including Richard Wagner's opera "The Flying Dutchman." The legend's haunting allure and themes of eternal punishment, redemption, and the relentless sea have captivated the imaginations of writers, poets, and playwrights for centuries. Artistic interpretations through paintings, illustrations, and sculptures have immortalized the ghostly vessel's image, portraying it navigating stormy seas, veiled in ethereal mist.

● Image 68: Visual Material ●

Beyond its maritime context, The Flying Dutchman has taken on allegorical significance, representing various themes in cultural narratives. The doomed captain and his ghostly ship have been interpreted as metaphors for human fate, eternal restlessness, and the struggle against destiny. The legend's resonance lies in its ability to symbolize the human condition, confronting mortality and the eternal quest for redemption or salvation.

The legend's cultural legacy extends to modern times, permeating popular culture through music, film, and entertainment.

Contemporary adaptations in cinema and music further amplify the legend's enduring appeal. Films featuring spectral ships, echoing the haunting presence of The Flying Dutchman, continue to captivate audiences, preserving the timeless allure of the legendary ghost ship.

The enduring impact of The Flying Dutchman on maritime lore is profound. Sailors and seafarers, steeped in tradition and superstition, have embraced the legend as a cautionary tale of the perils of the sea. Its portrayal as a harbinger of misfortune or as a symbol of doomed voyages resonates within maritime communities, adding a layer of mystique to seafaring folklore.

The enigmatic tale of The Flying Dutchman persists as a captivating maritime mystery, blending historical accounts, folk narratives, and scientific conjecture. As science and folklore interweave, the legend continues to intrigue, leaving an indelible mark on maritime heritage and cultural imagination, perpetuating the fascination with this ghostly apparition that still roams the oceans of human curiosity.

THE MYSTIQUE OF SHAMBHALA

Shrouded in mysticism and folklore, Shambhala has emerged as a legendary realm of spiritual enlightenment and utopia. Revered in diverse cultural narratives and esoteric teachings, the allure of Shambhala transcends geographical boundaries, captivating scholars, explorers, and mystics alike. This scientific article endeavors to unravel the enigmatic mystique of Shambhala by interweaving scientific inquiry, historical research, and esoteric traditions.

The exploration of Shambhala's mystique intertwines scientific inquiry with historical accounts, seeking to demystify the origins of this legendary realm. Scientific discourse delves into geographical hypotheses surrounding Shambhala's location, often posited within the Himalayas or Central Asia. Scholars and explorers draw connections between ancient texts and geographical anomalies, uncovering potential links to hidden valleys or remote regions that resonate with descriptions of Shambhala. Archaeological research, though inconclusive, has unearthed artifacts and manuscripts hinting at the existence of such a sanctuary.

• Image 69: Visual Material •

The mystique of Shambhala is deeply ingrained in ancient texts and religious scriptures, particularly prominent within Tibetan Buddhism. The Kalachakra Tantra, revered as a foundational text, vividly depicts Shambhala as a utopian land guarded by benevolent beings and envisioned as a haven of enlightenment and spiritual evolution. Historical accounts embedded in Tibetan Buddhist traditions narrate tales of great kings and spiritual masters guiding humanity toward higher consciousness within Shambhala's sacred precincts.

The folklore and cultural narratives enveloping Shambhala traverse diverse cultures, echoing tales of a fabled realm steeped in spirituality and serenity.   Shambhala, known by varied names like Shangri-La, Shamballa, or the Kingdom of Agharta, manifests in myths and legends across cultures spanning Asia, Europe, and beyond. These narratives embellish tales of hidden valleys, mystical gates, or sacred realms, preserving a consistent theme—a utopian sanctuary untouched by time, fostering peace, and spiritual evolution. The pervasive nature of Shambhala within global folklore underscores its universal appeal and enduring resonance.

Beyond its literal interpretation, Shambhala symbolizes ideals of enlightenment, harmony, and societal transformation. Its depiction as an idyllic paradise signifies the human quest for inner peace, spiritual awakening, and societal utopia. The allegorical significance of Shambhala resonates within cultural narratives, inspiring seekers of inner wisdom and

By blending scientific perspectives with rich historical accounts and tracing the echoes of Shambhala across diverse folklore and cultural narratives, the enigma of this mythical realm emerges as a compelling amalgamation of spiritual yearning and historical mystique.

• Image 70: Visual Material •

Within esoteric teachings, Shambhala represents an inner realm of consciousness—a metaphorical sanctuary embodying spiritual enlightenment and peace. Ancient prophecies envision Shambhala as a spiritual force transcending physical boundaries, fostering the evolution of human consciousness and societal harmony. This mystical interpretation resonates within various spiritual practices, inspiring seekers of inner wisdom and enlightenment.

Explorations seeking to unveil the existence of Shambhala have navigated the intersection of scientific inquiry, geographical expeditions, and the lasting impact of this mystical realm on global culture.

Throughout history, numerous expeditions have sought to locate the physical manifestation of Shambhala within the Himalayas or Central Asia. Utilizing remote sensing technologies and geographical surveys, explorers have scoured remote regions, aiming to validate the existence of hidden valleys or sacred landscapes that align with the descriptions of Shambhala. Despite these efforts, the elusive nature of the realm has thwarted empirical confirmation.

While empirical evidence of Shambhala remains elusive, the realm's mystique exerts a profound cultural impact. Shambhala's influence permeates art, literature, and philosophical discourse, serving as a wellspring of inspiration for creative expressions worldwide. Artists, writers, and thinkers draw upon its symbolism, portraying Shambhala as a metaphor for inner peace, spiritual enlightenment, and societal harmony. Its enduring legacy transcends cultural boundaries, resonating deeply within human consciousness.

• Image 71: Visual Material •

Beyond scientific exploration, Shambhala embodies a spiritual quest, transcending physical confines and tapping into the interconnectedness of human consciousness. Its depiction as a spiritual sanctuary fosters contemplation on societal well-being, unity, and the pursuit of inner wisdom. The enduring allure of Shambhala within spiritual circles underscores its relevance in humanity's collective quest for enlightenment and societal transformation.

The impact of Shambhala extends beyond a mere search for a physical location, shaping cultural discourse on spirituality, harmony, and the interconnectedness of humanity and nature. Its depiction as an aspirational realm continues to inspire introspection, fostering a deeper understanding of the human condition and our yearning for transcendence.

The enigma of Shambhala, rooted in a blend of scientific inquiry, historical accounts, folklore, and spiritual teachings, remains an enduring mystery that transcends empirical validation. While scientific exploration seeks tangible evidence, the mystique of Shambhala persists as a symbol of spiritual aspiration, enlightenment, and societal harmony. Its legacy continues to inspire seekers of wisdom, fostering a deeper understanding of the interconnectedness between human consciousness and transcendent realms.

Engaging in the exploration of cognitive marvels, we embark on an intriguing journey encompassing the uncanny Déjà vu experience, the profound impact of the Placebo Effect, and the elusive enigma surrounding the Mystery of Consciousness. These mind-bending phenomena beckon us to navigate the intricate labyrinths of the mind, where perception intertwines with reality, challenging our comprehension of the human experience.

"Mind-bending phenomena are the reminders that the universe is far more mysterious and complex than we can comprehend"

Michio Kaku

DÉJÀ VU - THE PUZZLE OF FAMILIARITY

As we journey through the labyrinthine landscape of the human mind, we encounter phenomena that defy easy explanation. Among these enigmatic experiences stands Déjà vu—the unsettling sense of familiarity in a moment that defies logical memory recall.

Our exploration of cognition, perception, and memory mechanisms has equipped us to grapple with the puzzle of Déjà vu. We've observed how perception molds our reality, stitching together sensory inputs to craft a coherent narrative. However, Déjà vu disrupts this narrative by introducing a sensation of reliving a moment without a clear memory trace.

● Image 72: Visual Material ●

Memory, as we've uncovered, is a complex and malleable construct. It isn't a perfect replica but subject to distortion and reconstruction. Déjà vu amplifies this complexity by presenting fragments of memory without a discernible source, challenging the reliability of our recollections.

While the phenomenon of Déjà vu remains complex and multifaceted, neuroscience has made significant strides in uncovering potential neural underpinnings through various studies and imaging techniques.

- *Temporal Lobe Involvement:* Studies utilizing functional MRI (fMRI) and EEG (electroencephalography) have indicated heightened activity in the temporal lobe during Déjà vu experiences. Dr. Anne Cleary, a cognitive psychologist at Colorado State University, explains that the temporal lobe, specifically the medial temporal lobe structures like the hippocampus and adjacent cortices, plays a vital role in familiarity and recognition processes. The activation of these regions suggests their involvement in the sudden sense of recognition or familiarity without an identifiable source.

- *Hippocampal Significance:* Research led by Dr. Akira O'Connor at the University of St Andrews in Scotland highlights the hippocampus's role in the Déjà vu phenomenon. Studies have shown that patients with damage or dysfunction in the hippocampus have a higher frequency of experiencing Déjà vu. This suggests that the hippocampus, known for its involvement in memory formation and consolidation, might contribute to the retrieval of familiar feelings without explicit memory recall.

- *Frontal Cortex Contribution:* Dr. Alan Brown, a professor of psychology and neuroscience at Southern Methodist University, emphasizes the involvement of the prefrontal cortex in Déjà vu. This brain region, responsible for higher cognitive functions, may contribute to the interpretation and evaluation of familiarity signals. Disruptions in the prefrontal cortex's function might lead to the misattribution of familiarity, contributing to the experience of Déjà vu.

● Image 73: Visual Material ●

Additionally, collaborative studies, such as those led by Dr. Christine Wells at Sheffield Hallam University, have employed neuroimaging techniques to examine brain activity patterns during Déjà vu episodes. These investigations have consistently revealed increased activity in brain regions associated with memory retrieval and familiarity processing.

Psychological theories regarding Déjà vu encompass a range of explanations that revolve around memory errors, attentional lapses, and the brain's attempt to reconcile similarities between different situations. These theories provide valuable insights into how the mind processes familiarity amidst new experiences.

• *Memory Errors and Associative Activation:* According to Dr. Vernon Neppe, a neuropsychiatrist known for his work on Déjà vu, memory errors may occur due to misfirings or rapid firings in the brain, leading to the activation of memories or information that might not correspond to the current experience. This theory suggests that the brain mistakenly retrieves fragments of memories, creating a sense of familiarity where none should exist.

• Image 74: Visual Material •

• *Lapses in Attention and Perception:* Dr. Anne M. Cleary's research at Colorado State University emphasizes the role of attentional lapses in Déjà vu. She suggests that moments of distraction or divided attention might disrupt the brain's ability to process incoming information accurately. This interruption in attentional focus could cause the brain to briefly misinterpret a new experience as a familiar one, resulting in a Déjà vu sensation.

• *Mismatch Resolution and Similarity Processing:* Dr. Chris Moulin, a cognitive neuropsychologist at the University of Grenoble Alpes, delves into the brain's challenge in reconciling similar but distinct situations. His research suggests that Déjà vu might occur when the brain encounters a situation that closely resembles a stored memory but isn't an exact match. The brain grapples with this discrepancy, leading to a feeling of familiarity without a clear memory trace.

Studies examining these theories often involve experimental setups to induce Déjà vu-like experiences in controlled settings. For instance, Dr. Alan S. Brown, a professor of psychology and neuroscience at Southern Methodist University, has

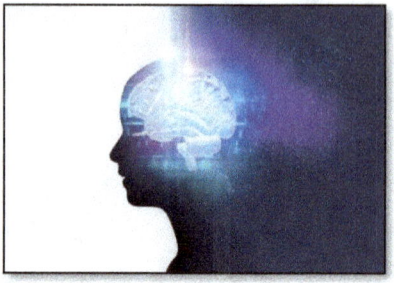

• Image 75: Visual Material •

conducted studies simulating Déjà vu in laboratory settings by presenting participants with word pairs that are semantically related but not explicitly shown together. These experiments aim to mimic the brain's potential associative activation and memory error mechanisms, shedding light on how Déjà vu-like experiences can be induced.

Yet, Déjà vu remains an enigma, inviting us to contemplate the intricacies of human cognition. Its occurrence challenges our current understanding, beckoning for further investigation into the intersections of perception, memory, and consciousness.

THE PLACEBO EFFECT: THE MIND'S HEALING POWER

In our exploration of the complexities of the human mind, one phenomenon stands out among the enigmatic demonstrations of the mind's influence: the placebo effect. This intriguing occurrence challenges our understanding of how belief, expectation, and the mind's power can profoundly impact health and healing.

As we've traversed the intricacies of perception, memory, and cognitive processes, the placebo effect emerges as a testament to the mind's remarkable ability to shape physiological outcomes. It hinges on the power of belief and expectation, showcasing how the mind's anticipation of a treatment's efficacy can yield tangible improvements in health. The intertwining of cognitive processes within the placebo effect has been a subject of extensive scientific inquiry, shedding light on how beliefs and expectations can lead to real physiological changes.

- *Association Formation and Memory:* Studies, such as those conducted by Dr. Fabrizio Benedetti, a neuroscientist at the University of Turin Medical School, have delved into the role of memory in the placebo effect. Benedetti's research demonstrates that individuals can form strong associations between a treatment and an anticipated outcome based on previous experiences or suggestions from healthcare providers. These associations are stored in memory and can influence future responses to similar treatments, even if they are inert.

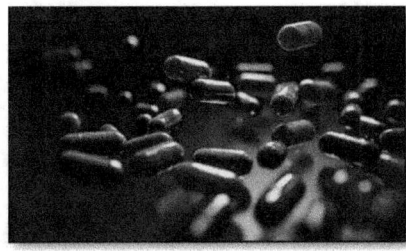

- Image 76: Visual Material •

- *Attention and Expectation:* Dr. Tor Wager, a neuroscientist at Dartmouth College, emphasizes the role of attention and expectation in the placebo response. His research, utilizing brain imaging techniques like fMRI, has revealed that when individuals expect a treatment to be effective, attentional processes are engaged, influencing how the brain processes pain and other sensations. This attentional focus amplifies the placebo effect by directing the brain's resources toward interpreting the expected outcome.

- *Learning and Conditioning:* Dr. Karin Jensen, a neuroscientist at the Karolinska Institute, explores the role of learning and conditioning in the placebo effect. Her studies show that through associative learning processes, individuals can develop conditioned responses to inert treatments. These learned responses, akin to Pavlovian conditioning, result in the activation of neural pathways that modulate physiological functions, such as pain perception or immune responses.

Research in these areas illustrates how the placebo effect is intricately woven into cognitive processes. The mind's ability to form associations, based on past experiences or suggestions, primes the brain to respond to a placebo treatment as if it were therapeutically active. This process involves the activation of neural circuits associated with memory retrieval, attentional focus, and learned responses, ultimately leading to physiological changes despite the inert nature of the treatment itself.

The placebo effect has far-reaching implications in healthcare, prompting a reevaluation of conventional medical paradigms. It highlights the significance of patient beliefs and expectations in influencing treatment outcomes. Moreover, neuroscience has unveiled insights into the neural mechanisms underpinning the placebo effect, revealing the involvement of brain regions associated with expectation and modulation of physiological responses.

The ethical considerations surrounding the use of placebos in medical practice have been a subject of ethical discourse and research, highlighting the delicate balance between potential benefits and ethical concerns.

• Image 77: Visual Material •

Informed Consent and Ethical Dilemmas: Dr. Franklin G. Miller, a bioethicist at the National Institutes of Health (NIH), has extensively examined the ethical implications of using placebos in clinical settings. He emphasizes the fundamental principle of informed consent in medical ethics, arguing that administering placebos without explicit disclosure and consent from patients raises ethical concerns. This practice may infringe upon patients' autonomy and right to make informed decisions about their healthcare.

Dr. Jeremy Howick, a philosopher of medicine at the University of Oxford, highlights the importance of maintaining trust and transparency in the patient-physician relationship. His research emphasizes that deceiving patients by administering placebos without disclosure could undermine trust in healthcare providers and the medical system as a whole. Preserving patient trust is crucial for effective healthcare delivery and patient outcomes.

The American Medical Association (AMA) and other medical organizations have established ethical guidelines emphasizing the necessity of informed consent in medical practice. According to the AMA's Code of Medical Ethics, physicians should provide patients with adequate information about treatments, including potential risks and benefits, to enable informed decision-making. This ethical stance aims to uphold patients' autonomy and respect for their values and preferences. Dr. Ted J. Kaptchuk, a researcher at Harvard Medical School, advocates for harnessing the placebo effect ethically. His work focuses on optimizing placebo responses through open-label placebo treatments that involve transparently providing patients with placebos and explaining their potential effects. Such approaches aim to utilize the mind's healing potential without resorting to deception, aligning with ethical principles of honesty and transparency in healthcare.

Ultimately, the placebo effect serves as a profound reminder of the mind's immense influence on health and healing. It urges us to explore further the interconnectedness of the mind and body, inspiring a deeper understanding of the human mind's transformative potential in promoting well-being.

THE MYSTERY OF CONSCIOUSNESS

Amidst our exploration of the intricate landscape of the mind, one enigma persists at the heart of our understanding: the mystery of consciousness. This captivating phenomenon transcends the realms of perception, memory, and cognitive processes, challenging us to fathom the essence of subjective experience and awareness.

In our intricate exploration of the human mind's complexities, one enigmatic phenomenon stands out as the ultimate mystery: consciousness. Despite our thorough examination of perception, memory's intricacies, the placebo effect's influence on health, and the interconnectedness of mind and body, consciousness remains an elusive puzzle that defies easy comprehension.

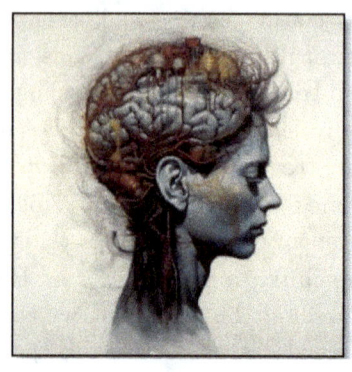

Consciousness embodies the entirety of our subjective experiences—our sensations, thoughts, emotions, and self-awareness. However, understanding how consciousness arises from neural processes and gives birth to our rich inner worlds presents an enduring puzzle, often termed the "hard problem" within scientific inquiry.

● Image 78: Visual Material ●

Our journey through cognitive processes and neuroscience has illuminated the inner workings of the brain. Yet, consciousness, distinct from brain activity, poses a philosophical and scientific challenge. Bridging the gap between neural processes and subjective experiences remains an enigmatic quest.

The quest to comprehend consciousness has led to the development of several prominent theories, each offering valuable insights into the nature of consciousness while acknowledging the inherent complexities.

Giulio Tononi, a prominent neuroscientist, proposed the Integrated Information Theory (IIT), positing that consciousness arises from the integration of information within neural networks. According to IIT, consciousness is correlated with the extent of information integration among interconnected neurons. Tononi argues that the level of consciousness corresponds to the amount of integrated information within a system.

In his studies, Tononi and his collaborators have employed mathematical frameworks to quantify the "phi" value, which measures the degree of integrated information in a system. While IIT provides a mathematical approach to understanding consciousness and has gained attention for its attempt to offer a quantitative measure, it also faces criticism for its complexity and challenges in experimental validation.

• Image 79: Visual Material •

Bernard Baars, a psychologist, proposed the Global Workspace Theory (GWT), which suggests that consciousness emerges from the brain's ability to broadcast information across various specialized neural modules or "workspaces." According to GWT, only information that gains access to this global workspace becomes consciously perceived and influences behavior. Baars's theory is supported by experimental evidence from cognitive psychology and neuroscience, revealing patterns of brain activity associated with conscious perception and decision-making. However, while GWT provides a framework for understanding the mechanisms of consciousness, it has been criticized for oversimplifying the complexities of conscious experience and the brain's distributed processing.

Despite the valuable insights offered by these theories, the quest to fully grasp consciousness remains elusive. Many scientists and philosophers acknowledge that no single theory provides a complete explanation of consciousness. The intricacies of subjective experience, the mind-brain relationship, and the emergence of consciousness from neural processes continue to pose challenges that transcend any singular explanatory framework.

The pursuit of understanding consciousness extends beyond scientific inquiry, encompassing philosophical contemplation. Philosophers delve into questions surrounding qualia, the nature of the self, and the mind-body relationship. The collaboration between science and philosophy underscores the multifaceted nature of consciousness but does not resolve its fundamental mystery.

As we conclude our exploration of the mind's mysteries, consciousness stands as an eternal enigma, urging humility in the face of the unknown. It invites us to embrace the limits of our understanding, acknowledging that the essence of consciousness may transcend current scientific and philosophical paradigms.

• Image 80: Visual Material •

Our journey through mind-bending phenomena leaves us at the precipice of an unending quest for knowledge. The mystery of consciousness beckons us forward, guiding our relentless pursuit of unraveling the profound depths of the human mind.

DREAMS - PORTALS TO THE SUBCONSCIOUS REALM

In the vast tapestry of human experience, dreams have remained an enigmatic and captivating phenomenon that transcends cultural boundaries. Across epochs and civilizations, these nocturnal reveries have stirred the human imagination, prompting wonder, exploration, and debate. Delving into the realm of dreams offers a journey into the subconscious, a landscape of infinite possibilities and intrigue.

Cultural perceptions of dreams have long been intertwined with spiritual beliefs, folklore, and societal customs across diverse civilizations.

In ancient Egypt, dreams held a profound significance and were considered sacred communications from the divine realm. Egyptians believed that dreams were messages sent by the gods or the deceased, offering guidance, predictions, or warnings about the future. Their interpretation of dreams was deeply intertwined with their religious practices and rituals. For instance, dream interpreters, known as "Masters of Secret Things," were revered for their ability to decipher the messages conveyed through dreams, often influencing important decisions, such as matters of state or personal destinies.

● Image 81: Visual Material ●

Similarly, ancient Greek civilization also placed immense importance on dreams. Greek mythology is replete with stories where dreams served as divine interventions or omens, providing insights into impending events or guiding heroes on their quests. Greeks believed that dreams were a medium through which the gods communicated with mortals, and temples often housed "sleep temples" where individuals sought prophetic dreams, aiming to receive guidance or healing.

Indigenous cultures across various continents have rich traditions and beliefs regarding dreams. Many indigenous communities perceive dreams as a bridge to the spirit world, enabling communication with ancestors or spirits. In some traditions, dreams are considered a source of wisdom and guidance, offering insights into resolving conflicts, making important decisions, or understanding natural phenomena.

In the realm of scientific inquiry into dreams, researchers have dedicated considerable effort to uncovering the mechanisms and functions that underpin these enigmatic nocturnal experiences. Utilizing sophisticated neuroscientific tools and methodologies, investigations into the nature of dreams have provided compelling insights, though they continue to present complex puzzles yet to be fully resolved.

Studies have revealed that dreams predominantly occur during the REM (Rapid Eye Movement) stage of sleep, a phase characterized by increased brain activity resembling wakefulness. During REM sleep, the brain engages in intricate processes, generating vivid, immersive dreamscapes replete with stories, emotions, and sensory perceptions. This phase is often associated with intense dreaming, marked by heightened neuronal firing patterns across various brain regions.

• Image 82: Visual Material •

Neuroscientific explanations propose that dreaming serves multifaceted functions within the realm of human cognition. One prominent theory suggests that dreams aid in memory consolidation, where the brain processes and consolidates newly acquired information from the day's experiences into long-term memory storage. Dreams may facilitate the organization and integration of memories, contributing to learning and knowledge retention.

Furthermore, dreams are considered instrumental in emotional regulation and processing. They provide a platform for the brain to navigate and comprehend emotions, allowing individuals to explore and confront subconscious feelings, anxieties, or unresolved conflicts in a safe, unconscious setting.

Beyond memory and emotional processing, some theories suggest that dreams contribute to problem-solving and creative ideation. They offer a realm where the mind can freely explore possibilities, synthesize novel ideas, and simulate scenarios that may aid in decision-making or innovation.

In the realm of psychology, Sigmund Freud's groundbreaking work emphasized the significance of dreams as gateways to the unconscious mind. His psychoanalytic theory suggested that dreams serve as a realm where repressed desires, fears, and unresolved conflicts manifest symbolically. Carl Jung expanded upon Freud's ideas, proposing that dreams also contain collective symbols and archetypes that reflect universal human experiences. Psychologists continue to explore dreams as a window into the psyche, aiding in self-discovery, healing, and understanding the complexities of human consciousness.

● Image 83: Visual Material ●

Dreams persist as an enduring frontier of human experience, inviting exploration and interpretation. As science continues to uncover the neurological underpinnings of dreaming, cultural diversity and psychological insights offer multifaceted perspectives on the significance and purpose of our nocturnal odysseys. They remain a testament to the enigmatic workings of the human mind, providing a gateway to the subconscious and inspiring wonder in our quest to comprehend their intricate mysteries.

Venturing into the cosmic unknown, we encounter the bewildering aftermath of The Tunguska Event, the tantalizing transmission of The Wow! Signal from Space, and the cosmic enigma embodied by Fast Radio Bursts. These celestial riddles, woven into the fabric of our cosmos, inspire us to unravel the cosmic puzzles that illuminate the vastness of our universe.

"Unexplained celestial events are the cosmic whispers that echo through space and time, inviting us to contemplate the secrets of the universe"

Brian Cox

THE TUNGUSKA EVENT - AN EXTRATERRESTRIAL MYSTERY

The Tunguska Event of 1908 stands as one of the most enigmatic cosmic occurrences in history, captivating scientific inquiry and folklore alike. On June 30th, 1908, a massive explosion rocked the remote Siberian region of Tunguska, devastating over 770 square miles of forested land. The event, equivalent to an estimated 10-15 megatons of TNT, resulted in the flattening of trees and the generation of seismic shockwaves felt thousands of miles away. Despite its colossal impact, the exact nature of the celestial body responsible for the explosion remains a scientific mystery.

• Image 84: Trees lie strewn across the Siberian countryside •

Local indigenous communities in the Tunguska region have passed down oral traditions that weave mystic narratives around the cataclysmic event. Tales of a "fire dragon" or a shamanic battle in the skies have been preserved through generations, echoing the awe and terror experienced by eyewitnesses of the Tunguska explosion.

According to these folk accounts, witnesses described a blindingly bright fireball streaking across the sky, followed by an earth-shaking blast that reverberated across the land. The intense light and thunderous roar terrified locals, who attributed the event to supernatural forces or celestial battles between cosmic beings.

These narratives, ingrained in the cultural fabric of the indigenous people, evoke a sense of wonder and mystery surrounding the Tunguska Event. The vivid retelling of the explosion underscores the profound impact it had on the collective consciousness of those who witnessed its awe-inspiring and terrifying spectacle.

In the wake of the Tunguska Event, numerous scientific expeditions and investigations sought to unravel the mystery behind this colossal cosmic occurrence. Scientists, geologists, and researchers from around the world embarked on missions to study the aftermath of the explosion and uncover its underlying causes.

• Image 85: Present tense •

One striking observation was the absence of a discernible impact crater, despite the magnitude of the devastation. This absence puzzled scientists, leading to the formulation of theories that postulated the celestial body responsible for the explosion disintegrated or exploded in the Earth's atmosphere before reaching the ground. The lack of substantial meteorite fragments added complexity to the investigation.

Examinations of the affected area revealed a radial pattern of fallen trees extending outward from the epicenter, indicating a shockwave originating from above rather than an impact from the ground. Researchers collected soil samples and tree core samples from the blast area, detecting anomalous levels of elements such as nickel, iridium, and other meteoritic materials. These findings provided indirect evidence of extraterrestrial origins, supporting the theory of an airburst event caused by a meteor or comet.

Furthermore, advancements in simulation models and trajectory analysis suggested that the object likely approached the Earth at a shallow angle, resulting in its explosive breakup at an altitude of about 3 to 6 miles above the ground. This high-altitude explosion generated a powerful blast wave that caused widespread devastation across the Siberian landscape.

Despite decades of scientific inquiry, definitive proof of the specific nature and composition of the celestial body that caused the Tunguska Event remains elusive. The absence of substantial meteorite fragments and the challenging terrain of the Tunguska region have hindered comprehensive investigations.

The search for remnants of the Tunguska meteorite has been hindered by several challenges, rendering its discovery and excavation an arduous task at present.

• Image 86: Crash site (after the event) •

• *Scarcity of Obvious Fragments:* The absence of distinct remnants poses a significant challenge. The meteorite's explosion extensively fragmented it, potentially causing complete disintegration in the atmosphere or burying the remnants deep underground. This complicates the search for fragments, making it a complex and demanding endeavor.

• *Vast Territory and Complex Conditions:* The Tunguska region comprises dense forests and marshlands, rendering search expeditions intricate. The affected area spans over 2000 square kilometers, and even with advanced technologies, surveying the entire expanse remains an immense undertaking.

• *Geological Challenges:* Locating the meteorite within the Siberian taiga and swamps poses geological challenges. Thick forests and the topsoil cover could obscure meteorite fragments, making their detection exceedingly difficult.

• *Temporal Changes and Material Decay:* The event occurred over a century ago, subjecting potential remnants to natural processes like soil erosion, forest fires, and biological degradation. These factors might have significantly altered the landscape and led to the decay or dispersion of the meteorite material.

• *Lack of Specific Discovery Locations:* The precise impact site of the meteorite remains uncertain. Despite researchers' efforts, exact coordinates or the impact area continue to be subjects of debate, complicating targeted search efforts.

These challenges collectively present formidable obstacles to uncovering remnants of the Tunguska meteorite. Intensive research and innovative technologies may prove essential for successful excavation and the definitive resolution of this enigma from the past.

The Tunguska Event remains an enduring astronomical mystery that continues to captivate scientific curiosity and cultural fascination. While scientific investigations have provided valuable insights into the event's characteristics and potential origins, conclusive evidence regarding the precise nature of the celestial body responsible for the explosion remains an ongoing pursuit.

• Image 87: The epicenter of the Tunguska event in Siberia •

The melding of local folklore with scientific inquiry underscores the magnitude and impact of the Tunguska Event on both the physical landscape and the collective memory of those who witnessed its awe-inspiring spectacle. As researchers persist in their quest for answers, the Tunguska Event stands as a testament to the enduring enigmas of our universe, inviting continued exploration and scientific scrutiny.

THE ENIGMATIC WOW! SIGNAL: ECHOES FROM COSMIC DEPTHS

Our exploration into unexplained celestial events has led us to encounters that challenge our comprehension of the universe. Among these mysteries is the captivating enigma known as the Wow! Signal—a signal from space that captivated astronomers and ignited imaginations, leaving an indelible mark on humanity's quest to decipher extraterrestrial communication.

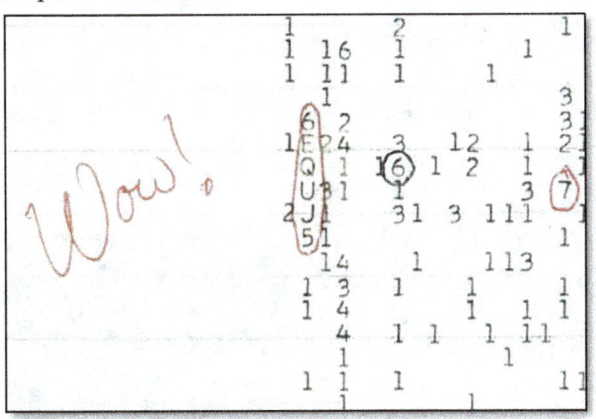

● Image 88: A color scan of the original computer printout of the "Wow!" signal ●

Building upon our exploration of consciousness, the placebo effect's influence on health, and the Tunguska event's cosmic puzzle, we venture into the depths of space in pursuit of understanding signals that transcend our known cosmic phenomena.

On August 15, 1977, the Big Ear radio telescope at Ohio State University captured an extraordinary signal originating from the constellation Sagittarius. The signal lasted for 72 seconds and exhibited characteristics suggesting a potential extraterrestrial origin. Astronomer Jerry Ehman, analyzing the data, scribbled "Wow!" beside the signal, signifying his astonishment.

The signal's distinctive feature was its narrowband frequency, falling within the hydrogen line—a range considered significant for potential interstellar communication. Despite subsequent efforts to detect a repetition or source, the Wow! Signal has never been detected again, leaving its origin and nature shrouded in mystery.

Scientists and astronomers have grappled with the Wow! Signal, offering diverse hypotheses and perspectives. Some suggest a natural astronomical phenomenon or Earth-based interference, while others entertain the tantalizing possibility of an extraterrestrial origin—a deliberate communication from an advanced civilization.

Statements from Researchers:

- Dr. Jill Tarter, an astronomer known for her work in the Search for Extraterrestrial Intelligence (SETI), stated, "The Wow! Signal remains an unparalleled anomaly in our search for extraterrestrial intelligence. Its transient nature and lack of repetition make it an elusive enigma."

- Dr. Robert Gray, a radio astronomer, commented, "Despite exhaustive attempts, we have been unable to replicate or definitively attribute the Wow! Signal to a natural or terrestrial source. Its potential

• Image 89: Visual Material •

extraterrestrial origin remains an enticing possibility."

The enigmatic nature of the Wow! Signal resonates deeply within humanity's quest for cosmic understanding, eliciting profound existential questions that transcend our current comprehension of the universe. At its core, the signal evokes contemplation about our place in the vast cosmos and the potential existence of intelligent beings. Possible Answers and Speculations:

- *Terrestrial or Natural Origin:*

While the Wow! Signal's source remains unidentified, some posit that it could have originated from Earth-based interference or an unknown natural celestial phenomenon. However, extensive efforts have failed to replicate or attribute the signal to such mundane origins.

- *Extraterrestrial Communication:*

The signal's narrowband frequency within the hydrogen line—the frequency considered significant for potential interstellar communication—propels speculation about its extraterrestrial origin. Could it be an intentional transmission from an advanced extraterrestrial civilization attempting to communicate with us?

- *Unresolved Mystery:*

Despite the scientific and philosophical contemplations, the Wow! Signal remains an unsolved cosmic mystery. Its transient nature, lack of repetition, and inconclusive source have rendered it an enduring enigma, leaving room for speculation and continued exploration.

The Wow! Signal, much like consciousness and unexplained celestial events, propels humanity to embrace scientific curiosity and philosophical wonder. It challenges us to venture boldly into uncharted cosmic territories, not only in search of scientific explanations but also to probe the profound implications of our place in the cosmos and the potential existence of other intelligent beings beyond our world.

FAST RADIO BURSTS: COSMIC PUZZLES IN THE CELESTIAL SYMPHONY

Our exploration into unexplained celestial events has unfurled a tapestry of cosmic mysteries that continue to challenge our understanding of the universe. Among these enigmas stands the phenomenon of Fast Radio Bursts (FRBs)—transient and powerful cosmic signals that intrigue astronomers, echoing the complexity of the cosmic symphony we seek to decipher.

Drawing from our exploration of consciousness, the placebo effect's influence on health, the Tunguska event's cosmic puzzle, and the enigmatic Wow! Signal from space, we venture further into the celestial realm to uncover the enigmatic Fast Radio Bursts and their perplexing nature.

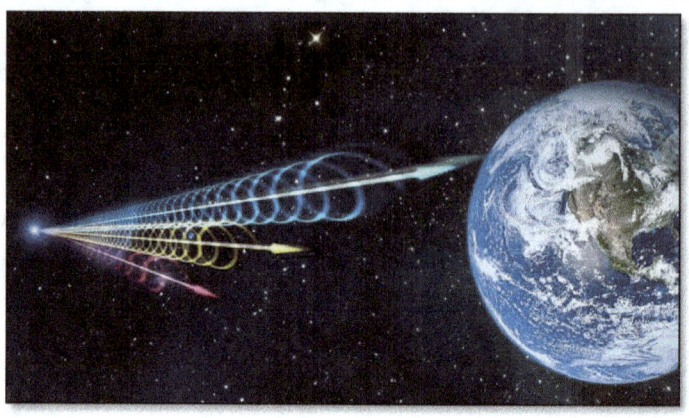

• Image 90: Visual Material •

Fast Radio Bursts are millisecond-duration radio signals of extragalactic origin, emitting as much energy in a few milliseconds as the Sun does in months. First discovered in 2007, these bursts, often one-time events, challenge astronomers due to their abrupt appearance, intense energy, and unknown sources.

- *Origins and Nature:*

The exact origins of Fast Radio Bursts remain elusive. Scientists debate potential sources, including neutron star mergers, magnetars, black holes, or highly energetic cosmic phenomena yet to be understood. These bursts, originating from billions of light-years away, arrive at Earth as faint echoes of cosmic events.

- *Scientific Inquiry and Speculations:*

Astronomers and researchers, employing radio telescopes and sophisticated detection systems, have captured dozens of Fast Radio Bursts. Despite this, the sporadic nature and lack of predictability hinder our understanding. Some speculate that these bursts could be cosmic cataclysms or phenomena involving exotic astrophysical processes.

Dr. Shriharsh Tendulkar, an astrophysicist, remarks, "The transient nature and immense energies of Fast Radio Bursts have left us puzzled. Identifying their origins is akin to deciphering a cryptic cosmic code that promises insights into the workings of the universe." Dr. Victoria Kaspi, a leading researcher in neutron stars and FRBs, notes, "FRBs represent a new frontier in astrophysics. Their puzzling nature fuels excitement as we strive to unlock the mysteries of these fleeting cosmic signals."

• Image 91: Visual Material •

Fast Radio Bursts, much like other celestial mysteries, beckon humanity to embark on scientific odysseys and philosophical contemplation. They challenge our comprehension of cosmic phenomena, invoking a sense of wonder and a yearning for discovery in the unexplored frontiers of the universe.

As we navigate the complexities of consciousness, grapple with unexplained celestial events, and ponder signals from cosmic depths, Fast Radio Bursts emerge as tantalizing puzzles in the grand cosmic tableau. Their transient nature and enigmatic origins stand as invitations for humanity to deepen our exploration, transcending the frontiers of knowledge as we seek to unravel the celestial mysteries that adorn the vast cosmic canvas.

CHAPTER 9: MODERN-DAY CONUNDRUMS

Journeying through the perplexing terrain of modern mysteries, we encounter the cryptic imprints of Crop Circles, the haunting reverberations of The Taos Hum, and the fiery enigma of Spontaneous Human Combustion. These contemporary enigmas, entwined with curiosity and intrigue, inspire us to seek answers beyond the ordinary, exploring the unexplained corners of our modern world

"Modern-day conundrums are the paradoxes that test the boundaries of our understanding, urging us to question, innovate, and adapt"

Michio Kaku

CROP CIRCLES: A TAPESTRY OF MYSTERY AND DEBATE

Crop circles, an intriguing phenomenon, stand as intricate patterns etched into fields of crops, igniting debates over their origins and significance. Crop circles, characterized by their intricate and precise geometric designs, have intrigued both the scientific community and the public, leading to varied theories surrounding their creation.

• Image 92: Stanton St Bernard, Wiltshire, 2009 •

- *Skeptic Perspective:*

Many skeptics dismiss crop circles as human-made creations, attributing their formation to talented individuals employing simple tools to flatten crops. They argue that the intricate patterns can be replicated using planks, ropes, and other basic instruments, indicating that no supernatural or extraterrestrial forces are involved. Joe Nickell, a prominent skeptic and investigator, notes, "Crop circles exhibit patterns that can be produced by skilled individuals using simple tools. The intricate designs, although visually stunning, are within the capabilities of human creators."

- *Meteorological and Environmental Factors:*

Alternative theories propose natural explanations, suggesting that meteorological phenomena or environmental factors might contribute to crop circle formation. Some researchers speculate that unusual wind patterns or localized weather events could cause swirling vortices capable of flattening crops in specific patterns. Dr.

Terence Meaden, a meteorologist and researcher, states, "I hypothesize that atmospheric phenomena such as vortices or whirlwinds may create crop circles. These weather-related events could cause the observed flattened crop formations."

- *Electromagnetic Forces:*

Other hypotheses focus on the potential influence of electromagnetic forces. Proponents suggest that microwave radiation or electromagnetic energy might influence crop growth patterns, leading to the creation of these intricate formations. Dr. Eltjo Haselhoff, a physicist specializing in crop circles, suggests, "Anomalous electromagnetic fields could potentially affect the growth of crops, resulting in the observed intricate patterns in crop circles."

• Image 93: Wiltshire, UK, 2001 •

- *Speculation about Extraterrestrial Origins:*

Enthusiasts and some researchers speculate about extraterrestrial involvement or attempts at communication with Earth. This theory proposes that these formations serve as intentional messages or signs from beings beyond our planet. Dr. Chandra Wickramasinghe, an astrobiologist, suggests, "Crop circles could be a form of communication from extraterrestrial intelligences attempting to convey messages or symbols to humanity."

- *Unknown Energy Sources:*

Some theorists propose involvement from unknown energy sources or forces not yet understood by mainstream science. They speculate that these energies might interact with the crops, resulting in the intricate formations observed in crop circles.

The debate rages on, with researchers and skeptics offering varying perspectives. While some physicists studying crop circles acknowledge the inexplicable intricacies and mathematical precision of certain formations, skeptics and investigators of paranormal phenomena, like Joe Nickell, maintain that crop circles are primarily hoaxes orchestrated by skilled individuals.

Crop circles serve as a canvas for human creativity, yet their unexplained formations raise questions about potential cosmic connections or symbolism. They symbolize modern-day enigmas that invite humanity to balance skepticism with open-minded inquiry. These formations represent the complexities of human perception, creative ingenuity, and the allure of cosmic mysteries yet to be fully grasped.

In conclusion, the debate surrounding crop circles intertwines with humanity's quest for understanding the unknown. As we navigate the web of theories and interpretations, crop circles symbolize the enigmatic intersections of art, science, and unexplored frontiers of human curiosity, prompting contemplation of the mysteries that lie both within and beyond our current understanding.

• Image 94: Belvoir Castle, 1991 •

THE TAOS HUM: ECHOES OF UNEXPLAINED ACOUSTIC MYSTERIES

The Taos Hum—this perplexing auditory phenomenon, much like the mysteries encountered before, challenges our understanding of unexplained phenomena and human perception.

The Taos Hum, a puzzling auditory phenomenon, presents as a persistent, low-frequency sound that only specific individuals in Taos, New Mexico, claim to perceive. Residents describe it as a low-level rumbling or humming, which has stirred curiosity and concern among locals and researchers alike for several decades.

Eyewitness accounts from affected individuals detail their experiences of hearing a continuous, low-pitched noise that seems to have no discernible external source. Descriptions vary, with some likening it to the distant sound of a diesel engine idling, a barely audible vibration, or a persistent low-frequency rumble.

• Image 95: The Taos Hum Visual

Residents who report hearing the Taos Hum often describe it as more than an occasional occurrence—it's an ongoing, sometimes intrusive presence in their lives. This auditory anomaly is typically perceived indoors, but some individuals claim to hear it regardless of their location or time of day, heightening their frustration and curiosity.

Scientific investigations into the Taos Hum have faced challenges in replicating or objectively measuring the sound due to its subjective nature. Studies aiming to record or analyze the phenomenon have struggled to capture consistent data, leading to a lack of conclusive evidence regarding its existence and characteristics.

Despite efforts to explain the Taos Hum through environmental, geological, or technological factors, no definitive cause has been established. Some theories suggest potential connections to geological resonances or atmospheric peculiarities specific to the region. However, these hypotheses remain speculative due to the lack of concrete scientific evidence linking these phenomena to the perceived sound.

Regarding the Taos Hum, some scientists explore explanations rooted in heightened auditory sensitivity or psychological factors. They propose that certain individuals might possess increased sensitivity to low-frequency sounds, allowing them to perceive noises that others may not notice. Alternatively, these scientists suggest that the phenomenon could potentially stem from psychological or physiological factors within the affected individuals, rather than from external environmental sources.

• Image 96: The Taos settlement •

Studies into auditory perception variations have revealed that some people possess a greater ability to detect and process low-frequency sounds. According to these studies, this heightened auditory sensitivity could lead some individuals to perceive sounds that fall within certain frequency ranges more acutely than the general population.

Psychological factors, such as stress or anxiety, have been theorized as potential contributors to auditory hallucinations or heightened perception of environmental sounds. Researchers postulate that stress-related conditions or psychological predispositions might influence how individuals interpret or perceive certain auditory stimuli.

Dr. Glen MacPherson, a researcher investigating mysterious hums and unusual auditory phenomena, has suggested that the perception of the Taos Hum might be a result of subjective auditory perception or psychological predispositions in certain individuals. He posits that the psychological state of the affected individuals could play a role in how they interpret or amplify environmental sounds.

Alternative hypotheses propose environmental or mechanical origins for The Taos Hum. Suspected sources include industrial machinery, electromagnetic interference, geological resonances, or atmospheric phenomena unique to the region. The incessant and mysterious nature of The Taos Hum has raised concerns about its psychological impact on affected individuals. Reports of insomnia, stress, and frustration among those who perceive the hum prompt further investigation into its potential effects on mental well-being. Local communities and researchers collaborate to study and address The Taos Hum's impact. Efforts include surveys, studies, and initiatives to understand its prevalence, impact, and potential mitigation strategies.

The Taos Hum joins the roster of unexplained mysteries that challenge our scientific understanding and perception. Much like our encounters with cosmic enigmas and enigmatic formations like crop circles, this auditory phenomenon serves as a reminder of the intricate tapestry of unexplained mysteries that invite humanity to delve deeper, combining scientific rigor with empathy and understanding. Its elusive nature fuels ongoing scientific inquiry and community engagement, urging us to explore the boundary between the known and the inexplicable.

SPONTANEOUS HUMAN COMBUSTION: ENIGMATIC FIRES WITHIN HUMAN BOUNDARIES

Spontaneous Human Combustion (SHC) is an enigmatic occurrence surrounded by debates and scientific skepticism, challenging our comprehension of unexplained fires that seemingly involve the human body.

As we venture deeper into the realm of inexplicable phenomena, SHC arises as a rare and controversial subject. Instances of individuals apparently bursting into flames without an apparent external ignition source have puzzled experts and the public alike for centuries.

Spontaneous Human Combustion (SHC) has sparked diverse theories attempting to decode this perplexing phenomenon. Some hypotheses delve into internal factors within the human body, proposing that chemical reactions or the presence of flammable substances like alcohol or fatty tissues might initiate spontaneous combustion.

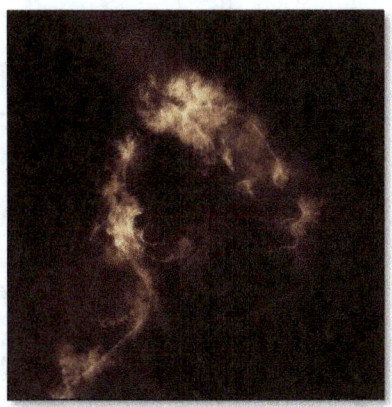

• Image 97: Visual Material •

However, the empirical evidence supporting these theories remains inconsistent and inconclusive. Eyewitness accounts and documented cases of SHC incidents often describe instances where the body, or parts of it, inexplicably burned without an apparent external ignition source. Some accounts portray scenes where the surroundings remain relatively unaffected, while the body shows signs of intense burning or complete incineration.

Scientific analyses of these cases often face challenges due to the lack of consistent patterns or identifiable causes. Investigations into the remains reveal localized burning, with body parts reduced to ashes, while nearby objects or surroundings remain unscathed.

Alternative hypotheses suggesting external ignition sources or accidental fires misinterpreted as spontaneous combustion propose overlooked external factors contributing to these incidents. These theories speculate that seemingly mysterious cases of SHC might actually involve conventional ignition sources, mistakenly perceived as spontaneous.

Eyewitness accounts and documented cases often highlight circumstances where the affected individual was found near potential sources of ignition. Some reports mention the presence of lighters, matches, or cigarettes in the vicinity, raising questions about accidental ignition rather than a spontaneous internal phenomenon.

In certain instances, investigations into alleged SHC cases revealed evidence of a potential external ignition source. Forensic examinations have occasionally identified burn patterns consistent with conventional fires ignited by external factors. Moreover, some cases displayed evidence of localized burning, suggesting a more plausible scenario of accidental ignition rather than the spontaneous ignition proposed by internal combustion theories.

• Image 98: Visual Material •

Forensic analyses of these incidents occasionally uncover burn patterns on the individual's clothing or immediate surroundings, indicating possible contact with external heat sources. Such observations challenge the notion of entirely spontaneous combustion and lean toward accidental ignition scenarios.

These alternative theories underscore the importance of considering external factors in alleged cases of spontaneous human combustion. While they do not discount the possibility of spontaneous ignition entirely, they highlight the need for comprehensive investigations that account for potential external sources of fire and contextual circumstances surrounding these incidents. The integration of such considerations might provide a more nuanced understanding of the perplexing phenomenon of SHC.

Forensic experts and scientists, while intrigued by the concept of SHC, remain cautious due to the scarcity of conclusive evidence supporting the notion of self-ignition solely from internal human processes. Dr. John DeHaan, a forensic fire scientist, underscores the need for further scientific inquiry, stating, "Spontaneous Human Combustion lacks empirical evidence supporting the idea of self-ignition from internal body processes."

SHC stands as an enigma that challenges conventional scientific understanding, raising profound questions about the delicate balance between internal biological processes and external environmental factors.

This mysterious phenomenon beckons us to tread the line between scientific inquiry and speculative reasoning. Its elusive nature calls for persistent scientific exploration, urging multidisciplinary efforts to unravel the complex puzzle it presents. As humanity grapples with the enigmatic nature of Spontaneous Human Combustion, it serves as a compelling reminder of the vast mysteries that continue to elude our comprehension within the intricate fabric of human existence.

THE CONTINUUM OF THE UNEXPLAINED

As we conclude this journey through the labyrinth of the inexplicable, it's clear that the world remains a canvas painted with mysteries waiting to be unraveled. From the enigmatic depths of lost civilizations to the unfathomable reaches of celestial events, humanity grapples with the boundaries of its understanding. The chapters navigated the immeasurable realms of history, science, culture, and the human mind, each revealing a shard of the vast mosaic that is the unexplained.

In our exploration of lost civilizations, we peered into the echoes of Atlantis, pondered the perplexing designs of the Nazca Lines, and pondered the haunting enigma of the Roanoke Colony's disappearance. These enigmatic chapters stand as testament to the fragility of history's threads and the mysteries buried within the sands of time.

The realm of cryptic creatures unveiled legends such as the elusive Bigfoot and the ever-enigmatic Loch Ness Monster, inviting contemplation on the blurred boundaries between myth and reality. Similarly, astounding archaeological discoveries like Stonehenge and the Moai of Easter Island whispered ancient secrets while leaving tantalizing puzzles unsolved.

Unsolved crimes and cold cases, such as the enduring riddles surrounding Jack the Ripper's reign of terror and the enigmatic disappearance of D.B. Cooper, persist as enigmatic specters in the labyrinth of history, defying resolution and entwining themselves within the fabric of our collective memory and fascination with the unknown. Additionally, the notorious Gardner Museum Heist, a brazen theft that robbed the world of invaluable artistic treasures, adds another layer to this mosaic of unresolved mysteries, leaving behind a haunting legacy that continues to intrigue and confound investigators and enthusiasts alike.

Science, our beacon of understanding, grapples with phenomena like the Bermuda Triangle and dark matter, compelling us to question the limitations of our current knowledge and encouraging the pursuit of further exploration and discovery.

Cultural enigmas and legends, from the cursed Hope Diamond to the spectral lore of the Flying Dutchman, thread through the tapestry of human culture, intertwining fact with myth and leaving indelible marks on our collective consciousness.

Mind-bending phenomena like déjà vu and the placebo effect tantalize us with glimpses into the frontiers of consciousness, urging contemplation on the intricate machinations of the human mind and perception. Moreover, delving into the mysterious landscape of dreams offers a portal to explore the depths of human consciousness, while unraveling the enigmatic nature of our waking reality. These links between dreams and consciousness prompt us to further explore the profound connections that underlie our cognitive experiences and subjective awareness. The cosmos itself, with events like the Tunguska explosion and the enigmatic Wow! Signal, beckons us to gaze upward and ponder the infinite wonders and mysteries that lie beyond our world.

Even in our modern era, conundrums like crop circles, the Taos Hum, and spontaneous human combustion stand as contemporary enigmas, defying easy explanation and beckoning further investigation.

As we reach the end of this voyage, one thing becomes abundantly clear: the pursuit of understanding the unexplained is not a journey with a final destination but an ongoing odyssey, a testament to the boundless curiosity and perseverance of the human spirit. Our quest to illuminate the darkness of the unknown will persist, fueled by wonder, skepticism, and the unyielding pursuit of truth. In embracing the unexplained, we embrace the very essence of our unrelenting thirst for knowledge, discovery, and the mysteries that continue to both confound and inspire us.

So, dear reader, as we close this chapter, let us not bid farewell to the mysteries, but instead, let us welcome them as companions on our journey, urging us ever forward into the uncharted territories of the unexplained.

The adventure continues.

AUTHOR'S REFLECTIONS: EXPLORING MYSTERIOUS REALMS

Penning down this book has been an incredible odyssey for me into the world of mysteries and enigmas that have always held an inexplicable allure. With each topic explored within these pages, I discovered new facets of mysteries that continued to captivate and astound me.

From ancient civilizations, steeped in the fog of legends and history, to the modern-day enigmas of science and culture - every theme was not just a challenge to my intellect but also an inspiration to delve into the depths of human nature.

It has been my long-standing desire to articulate these astonishing mysteries for a broader audience. I aimed not only to share my impressions and insights on each subject but also to open a door into the realm of unsolved riddles for anyone intrigued by this fascinating realm.

Through each page of this book, my hope is that readers will experience the same excitement and fascination that I encountered while exploring these remarkable topics. This is not just a book about mysteries and puzzles; it's an invitation to embark on an enthralling journey through the mysterious corners of human knowledge.

I extend my heartfelt gratitude to all the readers for joining me on this thrilling expedition. May this book serve as an inspiration for each individual to expand the horizons of their understanding and continue exploring the unfathomable mysteries that surround us.

With gratitude, *David Webb!*